T0321440

MIXING OF LIQUIDS BY MECHANICAL AGITATION

Chemical Engineering: Concepts and Reviews

A series edited by

Jaromir J. Ulbrecht
Center for Chemical Engineering
National Bureau of Standards
Washington, D.C.

Volume 1 **MIXING OF LIQUIDS BY MECHANICAL AGITATION**
Edited by Jaromir J. Ulbrecht and Gary K. Patterson

Volume 2 **DYNAMICS OF NONLINEAR SYSTEMS**
Edited by Vladimir Hlavacek

Additional volumes in preparation

ISSN: 0734-1644

MIXING OF LIQUIDS
BY MECHANICAL AGITATION

Edited by

JAROMIR J. ULBRECHT

Center for Chemical Engineering
National Bureau of Standards
Washington, D.C.

and

GARY K. PATTERSON

Department of Chemical Engineering
University of Arizona
Tucson, Arizona

GORDON AND BREACH SCIENCE PUBLISHERS
New York • London • Paris • Montreux • Tokyo

Gordon and Breach Science Publishers

P.O. Box 786
Cooper Station
New York, NY 10276
United States of America

P.O. Box 197
London WC2 9PX
England

58, rue Lhomond
75005 Paris
France

P.O. Box 161
1820 Montreux 2
Switzerland

14–9 Okubo 3-chome,
Shinjuku-ku,
Tokyo 160,
Japan

Library of Congress Cataloging in Publication Data
Main entry under title:

Mixing of liquids by mechanical agitation.

(Chemical engineering concepts and reviews, ISSN 0734-1644; v. 1)
1. Mixing. 2. Liquids. I. Ulbrecht, Jaromír J.,
1928–   II. Patterson, G. K. (Gary Kent), 1939–
III. Series.
TP156.M5M57   1985   660.2'84292   85-5253

CONTENTS

SERIES EDITOR'S INTRODUCTION

There has long been a need for monographs that discuss, in detail, specific subjects in chemical engineering. The aim of this series is to publish such monographs, each of which comprises articles developed from critical reviews of the relevant literature. The articles are written by experts whose contributions to the field under discussion are well known and significant. Articles written by more than one author still aim to present a single concept, although the exposition will certainly reflect the authors' somewhat differing opinions.

It is my belief that any review article worth reading must be not only informative, but also stimulating to the point of provocation. Thus, the reader will never see an "objective unbiased" review in this series (or, indeed, anywhere else—the mythical beast never has existed).

The authors are encouraged to structure their articles around their own and other closely related work. This approach, of course, involves the risk that the products will be noncritical self-serving summaries. It is to the credit of both the authors and the volume editors that no such thing has happened.

This first volume, *Mixing of Liquids by Mechanical Agitation*, will be followed by *Dynamics of Nonlinear Systems*. Further volumes are in preparation. Readers who have any suggestions for future topics for this series are respectfully urged to communicate them to the Series Editor.

J. J. Ulbrecht

PREFACE

This volume is a collection of studies showing a variety of approaches. Some parts of this book treat chiefly conceptual matters, while others are mainly concerned with design. The bulk of this book, however, represents an attempt to draw conclusions useful to the design of agitators through investigations of the mechanics of single- and multiphase flows in stirred tanks.

For most of us, not a single day will pass without involving us in some kind of mixing, be it stirring a cup of coffee in the morning or mixing drinks in the evening. And yet this ancient and omnipresent operation is very poorly understood. The design of mixers relies heavily on experience and engineering intuition based on common sense and a library of case histories (which, more often than not, is a well-guarded collection of *libri prohibiti*).

One of the difficulties in solving a mixing problem stems from the fact that, apart from the impossibly complex boundary conditions, the process is neither entirely random nor entirely deterministic. Thus, neither a purely analytical nor an integral "black box" approach is likely to provide a complete answer. Further, one does not stir a tank of liquid just for the sake of stirring. The ultimate aim is the promotion of a transport process or a chemical reaction in a single- or multiphase system. This precludes the establishment of either a single measure of successful mixing or of a uniform conceptual approach. Worse still, conflicting requirements having to do with, for instance, the dispersion of gas in a very viscous liquid, will result in designs making use of unorthodox geometries.

It should, therefore, be obvious that the variety of styles and approaches seen in this book has been brought about by the multifaceted nature of mixing (although the editors admit their share of responsibility for the colorful patchwork).

The first three chapters not only provide the necessary framework for a uniform treatment of mixing but also are meant to stimulate the development of fundamental concepts derived from underlying physical principles. The fourth chapter is the only one that deals with self-mixing and blending. The subject matter is, however, far from simple, since the liquids considered are rheologically complex.

The following four chapters deal with the dispersion of gases, liquids, and solids in liquids. Although the bulk of the material in these chapters deals with the fluid mechanics of multiphase systems, space is given, wherever appropriate, to the analysis of the interfacial area and mass transfer between the dispersed and continuous phases. This aspect of mixing is then treated to the fullest in the last chapter, which deals with scale-up procedures.

Nine authors participated in writing the nine chapters of this book. They are, in alphabetical order: Robert S. Brodkey, Pierre Carreau, Alvin W. Nienow, James Y. Oldshue, Gary K. Patterson, William E. Ranz, John M. Smith, Lawrence L. Tavlarides, and Jaromir J. Ulbrecht. They wrote about specialized areas in which they are experts due to many years of active research work. They recognize that, in many instances, their work would not have been possible without the financial support of various funding agencies, such as the National Science Foundation, the US Department of Energy, the British Science and Engineering Council, the Petroleum Fund of the American Chemical Society, and others. Their support is gratefully appreciated.

J. J. Ulbrecht
Gary K. Patterson

Chapter 1

Fluid Mechanical Mixing— Lamellar Description

WILLIAM E. RANZ

Department of Chemical Engineering and Materials Science, University of Minnesota, Minneapolis, MN 55455, USA

1. INTRODUCTION

It is not obvious how one should apply the mathematical apparatus of fluid mechanics to mixing. With facilities for massive computation one is tempted to solve the Navier–Stokes equation for particular boundary and initial conditions and for laminar motion of fluid volumes in a particular flow field. However, the result of such a time-consuming computation is as specific as a single physical experiment, and many such experiments lead only to empirical generalizations. With highly developed statistical theories of turbulence, one is also tempted to add the concepts of time-averaged concentration intensity and concentration scale and seek revelations from further manipulation of the equations of change using distributions of values of the averaged quantities. Unfortunately, in a structured continuum where mechanical mixing, interdiffusion of chemical species, and chemical reaction, can occur at differential space and time scales which range from the molecular scales of a solution phase to the macroscopic scales of a mixing device, spatial averaging is also required.

The purpose here is to promote the lamellar description of fluid mechanical mixing[1-5] as an essential part of any theory of mixing and as a prerequisite for explaining and predicting strange happenings when mechanical mixing, diffusion, and chemical reaction occur

1

simultaneously. In a real sense, the lamellar description is a means of reconciling the disparate points of view, deterministic or stochastic, which in the past have been proposed for analyses of mixing processes.

1.1.   Defining States of Mechanical Mixedness

A viscous liquid of constant mechanical properties, in which one can initially designate and continue to identify material volumes and intermaterial areas separating these volumes, is the idealized fluid for defining states of mechanical mixedness. There is no interfacial tension in the intermaterial areas. If diffusion of molecules occurs, they exchange identification when they cross the intermaterial area. It is self-evident that initially designated fluid volumes and boundary flows remain continuously connected in space and time. An instantaneous cross section of a mixed flow shows sectioning of intermaterial areas and these delineate laminae or striations. A marbleized structure, as in Fig. 1 will be seen in cross-sections.

If Fig. 1 represents a differential volume in a structured continuum and if its scale is large enough, one can define as a measure of mechanical mixedness the continuum quantity a_v, intermaterial area per unit volume, or the volume averaged striation thickness

Figure 1 Differential sections of a fluid during mixing. Originally designated material volumes have formed a lamellar structure, and a noticeable amount of diffusion has not yet occurred. A section does not give a direct measure of striation thickness s.

$s \equiv a_v^{-1}$. Curvature and folding of the lamina, also an important aspect of mechanical mixing, will occur at much larger scales and will not be seen in Fig. 1. A differential time scale for the same structured continuum can be designated as the time interval in which the striation picture shows little change, say $d(\ln s) < 10^{-1}$.

If the initial fluid volumes are miscible solvents characterized by differences in solute concentration, then parallel structures of concentration variation would be seen except that these could be modified by diffusion and chemical reaction. For example, a cook fixes such a laminar structure in a marble cake by baking it. Someone studying mixing of color-coded viscous liquids can freeze or gel the mixed state for sectioning. Striations will be seen if there has been little diffusive penetration of the coloring agent into adjacent striations and if the resolution scale of the observing instrument is finer than the striation thickness.

At least three kinds of structured continuum quantities are needed to describe the state of mechanical mixedness:

1. Mass or volume fraction of originally designated mass(es) or flow(s).
2. Striation thickness(es) of the designated mass(es).
3. Some measure of the distribution of values and ordering of striation thickness(es).

The last quantity presents much difficulty even with an idealized fluid to be mixed. Since local variations in density and viscosity may also be significant, a completely structured mechanical continuum confounds reason. We must work with a continuum "sufficiently structured" to explain phenomena we can anticipate and observe. Because it lends itself to analysis of diffusion and reaction in mixing flows, volume fraction of one kind of ideal flow of two kinds of fluid and striation thickness are chosen here as the minimum number of measures of mechanical mixedness. Some mention will be made of distributions of striation size, but this aspect of mixedness must wait for future development of lamellar descriptions.

1.2. Initial and Boundary Values

The topological results of mixing are dependent on the topological nature of the initial and boundary conditions, how one starts a batch mixer, how one feeds flows into a continuous mixer.

Flow separation and attachment lines are of first importance for a consideration of initial and boundary conditions in mixing. Unfortunately, such lines can move on a surface; but because they cannot move in principle, we shall pretend that such movements are not important. Furthermore, on the boundaries between a solid surface and a mixing flow, we have to introduce the additional concept of an intermaterial line of contact, a line which does not move relative to the surface. Thus, a continuous lamina and its intermaterial area has fixed attachments with all solid boundaries, moving and stationary, as well as all flow boundaries.

Examples (Fig. 2) illustrate the problem of assigning values of mixedness to initial and boundary conditions and of specifying the differential scale of a structured continuum:

1. A tank of diameter D contains liquid A on the bottom with depth s_A and liquid B at the top with depth s_B. There is an intermaterial area between the two designated fluids equal to $\pi D^2/4$. The volume fraction of A, ϵ_A, is equal to the ratio $s_A/(s_A + s_B)$. A defined average striation thickness for fluid A, could be the reciprocal of the intermaterial area per unit volume of fluid A, that is, $s_A \equiv (s_A \pi D^2/4)/(\pi D^2/4) = s_A$. The distribution of striations in this initial case would be designated by an $s_B \neq s_A$ and an A–B–A–B lamellar structure. Initially, the differential space scale of the arbitrary continuum is D, but this scale within the tank soon decreases with stirring.

   If a stirrer is now introduced, it should be obvious that at subsequent times times the measures of mixedness have large variations throughout the vessel space, even though one can arbitrarily define averages for the whole volume of fluid in the tank. Since the stirrer also causes an internal circulation and since the scale of striations imparted to the flow passing through the stirrer is related to some scale d associated with the stirrer, it is more reasonable to convert this batch system to a continuous flow system where the input flow may have been through the stirrer a varying number of times previously.

   At a later stage all of the intermaterial area would be attached to the original intermaterial perimeter. A lamina of A would coat the upper and side surfaces of the original A volume, and a lamina of B would coat the lower and bottom surfaces of the tank.

2. Flow w_A of fluid A is introduced concentrically through a pipe of diameter d into a flow w_B of fluid B in a larger pipe. The separation line, which is also an intermaterial line, has a peri-

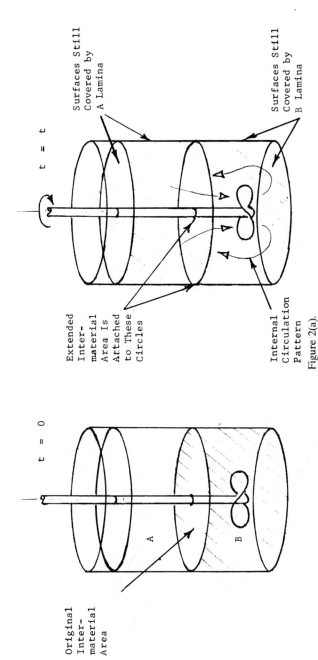

Figure 2 Examples of initial and boundary conditions for lamellar descriptions of fluid mechanical mixing. (a) Mixing of two liquids in a stirred tank. Initial conditions are clearly defined. Later boundary conditions are for internal flows which have been circulated many times through the stirrer pump. All intermaterial area extends from the original periphery of attachment. (b) Mixing of a laminar flow in a pipe; central and side-by-side contacting. All intermaterial area extends from the periphery or line of initial contact, even when the flow is turbulent downstream and the intermaterial area becomes a lamellar tangle. (c) Submerged jet mixing. All intermaterial area of laminae, observed as striations in an outflow cross section, extends to the stagnation point of the two jets and to the periphery of the delayed jet orifice.

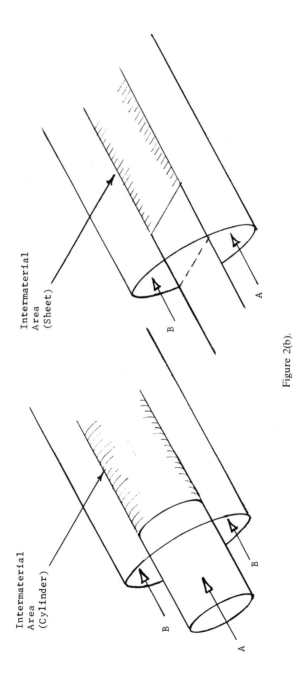

Figure 2(b).

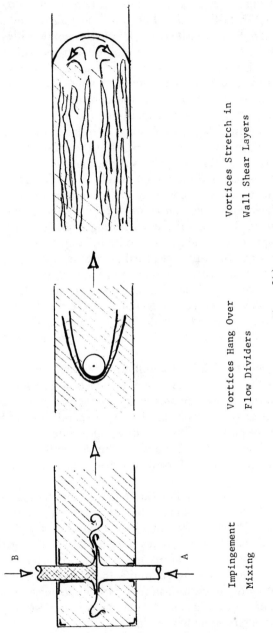

Impingement
Mixing

Vortices Hang Over
Flow Dividers

Vortices Stretch in
Wall Shear Layers

Figure 2(c).

meter πd at the inflow cross section. The inlet boundary value of intermaterial area per unit volume of flow A is $\pi d(w_A/\rho_A)/(\pi d^2/4) = 4/d$. For an averaged inflow, $s_A \equiv (d/4)((w_A/\rho_A) + (w_B/\rho_B))/(w_A/\rho_A)$. For constant densities and equal flows, $s_A = d/2$ by this method of defining striation scale. A many-tubed introduction of A, with a smaller value of d, would significantly decrease the boundary value of s_A. If tubes of various diameters were used, these would also determine the distribution of s_A in the boundary flow.

If equal flows of A and B were introduced on each side of a diameter D in pipe flow, $s_A = \pi D/4$ at the inflow boundary. The separating diameter would be an intermaterial line to which all intermaterial area downstream would be attached. At infinite time after flow has become steady, an intermaterial line would have attached by diffusion to the pipe wall and extended downstream from the ends of the separating diameter. In the case of center pipe introduction there would be no intermaterial lines on the pipe wall. In both cases further mixing does not occur in a straight pipe, in the sense that intermaterial area per unit flow volume increases or defined striation thickness decreases only when the flow cross section decreases. Here the differential space scale of the perceived structured continuum in the radial direction is D or d and in the axial direction is infinitesimal at steady flow.

3. Submerged jets of fluid A and fluid B impinge on one another in a jet mixer such as that used in reaction injection molding. Intermaterial area originates at the impact-stagnation-separation point and grows radially as the jets expand into a thin sheet. Each jet separates at its orifice, and intermaterial area is generated as in Part 2 because the jet entrains a recycled, relatively thin lamina of itself, which it joins, or of the opposite fluid.

Since the velocity of flow at radius r in the expanding sheet tends ideally to be the same as the jet velocity V_j, laminae formed by the primary flow have an $s(r) \geqslant d_j^2/8r$. This rapidly thinning, submerged sheet flow becomes unstable at small r of order $2d_j$, and the lamina roll up into vortices. Intermaterial area generation and lamina thinning continues thereafter only in regions of stretch of vortex lines, diffusion of vorticity, extension of the flow into runner and mold, and in new vortex motion generated by mixing configurations downstream.

In principle, assuming a leading jet wets the walls confining the flow, all intermaterial areas of laminae observed in an outflow cross section extend to the orifice periphery of the opposite jet or to the stagnation point of the two jets. If the differential scale of the structured continuum is taken as d_j and the differential time is taken as d_j/V_j, the inflow s could be specified as in Part 2. If the differential scale is taken as $s(r = 2d_j)$ and differential time as $s(r = 2d_j)/V_j$, then $s(r = 2d_j)$ is the boundary value for an inflow section arbitrarily set at $r = 2d_j$.

In all of the above examples, it should be apparent that a structured continuum is arbitrarily defined and that the definition has consequences with regard to usefulness of equations of change involving defined structured quantities. It is noted, especially for mixing, that initiating conditions are still needed for mixing flows that can only approach steady conditions assymptotically. A time dependent quantity, say $s(t)$, will always be needed to describe a mixing flow which otherwise can be classified as a steady flow.

1.3. Mixing Frame of Reference

Although favored for applications to practical steady-flow problems, Eulerian frames of reference are not the best frames for describing mixing, especially when diffusion and reaction are taken into account. Most steady flows which have been studied by fluid mechanists are not mixing flows. Consider, for example, steady viscous flow with two non-vanishing velocity components, that is, two-dimensional flows or flows with an axis of symmetry, can be expressed in terms of the stream function $\psi(x, y)$. Constant values of ψ from a flow boundary can represent the intermaterial areas between fluids being mixed, provided the values of viscosities and densities are the same for each fluid or appropriate averages can be specified. At small Reynolds numbers $\psi(x, y)$ is often obtained by assuming that the inertial terms of the Navier–Stokes equation are negligible. Such flows may have regions of higher shear, but there is as much unmixing as there is mixing between symmetrical flow boundaries. For example, consider Poiseuille flow in a tube or viscous flow normal to a cylinder. If normals of striations could enter such flows oriented parallel to the streamlines, mixing will occur; but such a boundary condition cannot be realized.

Consideration of the constraints on boundary conditions suggests that steady mixing flows must be three-dimensional, that geometric tricks which periodically reorient the striations are needed for mixing in two-dimensional flows, that shear and mixing are caused by fluid deformation but shear is not the cause of mixing, and that a Lagrangian history of differential mixing elements must be followed to understand the details of the mixing process.

At small scales in the Lagrangian frame, which translates with the mean motion, a flow can be resolved[6] into a pure straining motion, a dilational motion (if the fluid is compressible), and a rigid body rotation. The first two flows add to an axially symmetrical stagnation flow. If now a special "Lagrangian mixing frame" is oriented on the striation normal, then the flow in that special frame is also an axially symmetrical stagnation flow with velocities made up of appropriate components of the straining and dilational motions. Motion in the mixing frame depends on the instantaneous orientation angles between the striation normal and the principle directions of the strain. These angles, or equivalent vector "directors", are an additional type of structured continuum quantity needed to describe the local rate of mixing. They have to be established by following, in the Lagrangian frame, changes of the angles with time from specified values at the inlet flow boundaries or at initial time.

If the deformational-dilational frame is viewed at sufficiently small scale, continuity requires that flow in the frame is necessarily an elliptically symmetrical stagnation flow. The dilational aspect imposes a three-dimensional character to the flow, but an incompressible fluid tends to have a two-dimensional local flow. Such a two-dimensional flow exists in simple shear, in laminar flow, and in a non-stretching vortex. A two-dimensional tendency is observed in the microscales of turbulence. Since viscous dissipation of energy within any volume in the local flow is equal to viscous work on the volume and since the pressure field appears to be symmetrical to an observer in the volume, kinetic energy appears also to be conserved; and the flow is expected by mechanical laws to be two-dimensional.

There is much dynamic uniformity in a stagnation or stretching flow. The instantaneous flow field appears to be the same to every fluid particle in the flow. A straight material line remains straight and a material plane remains planar while increasing dimension in the direction of stretch. A material volume soon becomes a thinning slab, and the mixing frame approaches the same orientation as the deformational frame[7].

1.4. Time-dependent Quantity for Mixing Process

Assuming that the mixing frame is represented by a two-dimensional stagnation flow and that the x-coordinate is the direction of area stretch or contraction and the y-coordinate is the direction of striation thinning or thickening, the velocity field is given by

$$v_x = -\alpha x \qquad v_y = \alpha y \qquad v_z = 0 \qquad (1)$$

where $\alpha = \alpha(t)$. For a two-material incompressible fluid being mixed, time dependency can be assigned to any one of three quantities, α, a_v, or s, which are related as follows

$$\alpha = -\frac{d(\ln a_v)}{dt} = \frac{d(\ln s)}{dt} \qquad (2)$$

Because it is the quantity of most usefulness in applications to diffusion and reaction, $s(t)$ is the quantity of choice for describing the effects of mechanical mixing on mixing phenomena.

It should be noted that the value of s can be increasing (positive α) as well as decreasing (negative α). Mechanical mixing is reversible if the Lagrangian flow is reversible. In a practical mixer the value of α must be generally negative. Furthermore, it is worthwhile to note that in isotropic mixing in an isotropic turbulence field the time-averaged value of α would be zero. This means that turbulence does not mix when the stretch axis becomes uncorrelated with the normals of intermaterial areas.

2. APPLICATION OF LAMELLAR DESCRIPTION

2.1. Diffusion and Reaction in a Stretching Lamina

Since a fluid material containing chemical species A in concentration c_A assumes the form of a stretching or contracting lamina with nearly uniform conditions in planes parallel to the laminae, the differential equation of change of c_A in the Lagrangian frame fixed on the lamina takes the form

$$\frac{\partial c_A}{\partial t} + \alpha y \frac{\partial c_A}{\partial y} = D_A \frac{\partial^2 c_A}{\partial y^2} + R_A \qquad (3)$$

for an incompressible fluid in which there are relatively small concentrations of c_A with constant diffusivity D_A. If there is homogeneous chemical reaction, R_A is the volumetric generation rate of A. In a not uncommon case, $R_A = -kc_A$ for an apparent first-order reaction of A with some other reactant in large excess. A comparable equation can be written for temperature change in case hot and cold laminae are in contact or the chemical reaction produces significant sensible heat changes.

The convection term involving α and the reaction rate term R_A, when it is represented by a first-order reaction or when there is no reaction and $k = 0$, can be made to disappear with changes of variable that feature a warped time τ [7]:

$$Y_A \equiv \frac{c_A}{c_{A0}} \exp\left[kt(\tau)\right]$$

$$\xi \equiv y/s$$

$$\tau \equiv D_A \int_0^t s^{-2}\, dt'$$

The differential equation of change of Y_A thereby reduces to

$$\frac{\partial Y_A}{\partial \tau} = \frac{\partial^2 Y_A}{\partial \xi^2} \tag{4}$$

a familiar differential equation having solutions for a variety of initial and boundary conditions.

Thus, an elaborately complex system has been reduced to a level of rational analysis by describing convection, diffusion, and reaction in a Lagrangian frame of reference with one spatial and one time dimension, and by assigning all time independency to a single quantity whose value comes only from fluid mechanics.

2.2. Relationship of the Lamellar Description to Experimental Measures of Mixedness

A number of experimental measures of mixedness have been proposed and used in mixing studies. Frequency statistics of concentration fluctuations measured with a sufficiently fine probe in a Eulerian frame are the most direct measures of lamellar thickness distributions[8]. Frequency becomes scale through an averaged velocity; but scale is distorted by non-correlation with velocity fluctuations, and the sensing volume of the probe never seems to be small enough to record the smallest lamina. Statistics of the concentration fluctuations (intensities) are a measure of diffusive as well as mechanical mixing.

Measures involving simultaneous diffusion and reaction have also been used to compare the effectiveness of various types of mixing devices[3, 9–11]. To follow mechanical mixing alone these methods require considerable interpretation because they are diffusion and reaction specific. Here at some position in a flow, translated as mixing contact time, the space averaged extent of reaction, as measured by adiabatic temperature rise or color change, is proposed as a state of practical mixedness.

To show a closer relationship consider a case of mixing and diffusion without reaction ($k = 0$) when initially $c_A = 0$, where $-s_0/2 \leq y \leq 0$ and $c_A = c_{A0}$, where $0 \leq y \leq s_0/2$. At $t = 0$, alternating layers of thickness s_0 have concentration c_{A0} of species A. Subsequently the laminae are stretched, and $\partial c_A/\partial y = 0$ at $y = \pm s(t)/2$. In dimensionless variables and warped time, initial conditions at $\tau = 0$ are $Y_A = 0$, where $-0.5 \leq \xi \leq 0$ and $Y_A = 1$, where $0 \leq \xi \leq 0.5$ and boundary conditions are $\partial Y_A/\partial \xi = 0$ at $\xi = \pm 0.5$ for $\tau > 0$. Solution of the differential equation in warped time is[12]

$$Y_A(\xi, \tau) = 0.5 + \sum_{n=1}^{\infty} \frac{2 \sin (0.5n\pi) \cos [n\pi(\xi + 0.5)]}{n\pi} \exp (-n^2\pi^2\tau)$$

(5)

The concentration field $Y_A = c_A/c_{A0}$ is now given in terms of $\xi = y/s(t)$ and $\tau(t)$ for both molecular diffusion and the convection of laminae thinning if $s(t)$ is known for the flow situation. An initial flat-toothed concentration function $c_A(y)$ becomes at some later time a smoothed ripple function of small amplitude and short wavelength with an ultimate uniform $c_A = 0.5c_{A0}$ at infinite time. This result is a prediction of scale and intensity of segregation[13].

When τ is sufficiently large, the series of Eq. (5) is dominated by its first term, and the maximum deviation from final value $Y_A = 0.5$ will be approximately $(2/\pi) \exp (-\pi^2\tau)$. To diminish the maximum deviation to 0.05, that is, a deviation of less than 10% of the final mean, $\tau > 1/4$.

Thus, an indicator change in color at a measurable distance downstream in a multistaged mixing device can be interpreted as a measure of an average value of periodic α when $s(t = 0)$ and D_A are known.

When fast enough chemical reaction occurs, the rate of reaction is controlled by convection due to mechanical mixing and diffusion of reactant to a reaction plane which is parallel to the intermaterial area. Much more complex cases, where diffusion as well as chemical reaction rate in a reaction zone control overall rate, can be subjected to numerical analysis, but use of mixing-reaction experiments as measures of $s(t)$, D_A, or reaction rate constant k, is subject to caveat[14]. When mixing (convection and diffusion) is slower than chemical reaction, the constraints can be specified by:

$$\frac{kc_{B0}s_0^2}{D_A} \geqslant \frac{kc_{B0}s^2}{D_A} \equiv \phi > 10^4 \tag{6}$$

where c_{B0} is the initial concentration of reactant B in fluid B. When mixing and reaction rates are comparable,

$$1 < \phi < 10^4 \tag{7}$$

When slow chemical reactions control,

$$\frac{kc_{B0}s_0^2}{D_A} \geqslant \frac{kc_{B0}s^2}{D_A} \equiv \phi < 1 \tag{8}$$

Most mixed gas reactions, except combustion, are in this category.

A direct way to measure mechanical mixing only in terms of $s(t)$ is to mix non-diffusing, identifiable fluids, freeze the striations at time t, cross section the solids, and measure the striation thicknesses. Experimental difficulties in this method will be no worse than interpretative difficulties with methods involving significant diffusion and chemical reaction.

2.3. Fluid Mechanical Mixing in Simple Shear and Stagnation Flows

Figure 3 illustrates the fate of a striation initially oriented across the flow when a uniform shear is suddenly imposed. However, the figure does not show how to obtain the initial orientation in a continuous flow. Note that this kind of mixing is reversible[15] and that the measure α of the rate of mixing in the mixing frame is eventually inversely proportional to time after passing through a minimum value when the striation is oriented at 45° to the shear. Note also that the mixing frame, based on the striation, rotates at decreasing rate with time. After sufficient time at constant shear the mixing rate becomes negligible.

Figure 4 shows mixing in a stagnation flow. If the initial orientation of the striation s is in the y-direction, laminae are oriented in the most favorable way for maximum rate of thinning when $\alpha(t)$ is of negative value. Again the flow and fluid mechanical mixing are reversible. If the flow is reversed, the previous most favorable

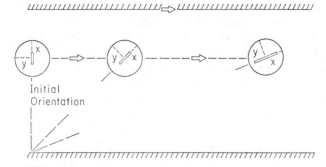

Figure 3   Stretch of fluid volume in shear flow. Mixing frame rotates at decreasing rate. In mixing frame $\alpha = -G(Gt)/(1 + (Gt)^2)$ where G is shear rate.

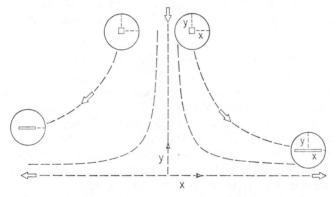

Figure 4   Stretch of fluid volume in stagnation flow. Mixing frame does not rotate. In mixing frame $v_x = -\alpha x$ and $v_y = \alpha y$ where $\alpha = \alpha(t)$.

orientation becomes most unfavorable, and the subject striation thickens rather than thins. In Fig. 4 the smaller scale mixing frame does not rotate with time; and the large scale stagnation flow, except for scale, is the same as the mixing frame flow.

Consideration of mixing in these two flows, as defined by s and $-d(\ln s)/dt$, demonstrates the speciousness of common statements made about fluid mechanical mixing; (a) mixing is caused by shear flow; (b) mixing is caused by turbulence; (c) mixing is proportional to mechanical energy dissipated by viscous action. All nouns capable of quantification in the foregoing statements are effects of fluid deformation caused by the geometry confining the flow and the initial and boundary conditions of the flow. Insofar as these quantities are effects of the same causes, and only the same causes, they can be related to one another on a one-to-one basis.

Since a time-dependent quantity and an initial geometric orientation are needed to describe the nature of fluid mechanical mixing in these two-dimensional flows, one should expect that continuous fluid mechanical mixing requires a three-dimensional flow or a geometric reorientation after a limited time in a two-dimensional flow of macroscopic scale. The fold after stretch in a taffy pull, or the one-quarter rotation during kneading of dough, is an intermittent geometrical reorientation of striations at a scale larger than striation scales and at times longer than differential thinning times.

2.4. Mixing Caused by Passage of a Stagnation Line Across an Intermaterial Area in a Wake Striation (Mixing action of a whisk, egg beater, prong kneader without vortex street in wake)

When an object with a stagnation line on its leading edge or face crosses an intermaterial area in a fluid, the intermaterial area is wrapped over the stagnation line like a blanket over a clothes line. Eventually, if the intermaterial area makes close enough contact with the stagnation surface, the blanket will dissolve and intermaterial lines will sweep over the surface of the object and leave it wetted only with the fluid it entered. Diffusional processes are involved in any realistic boundary condition, but here we assume a fluid mechanical intermaterial area which never breaks.

For a simple analysis of this mixing action, which is necessarily accompanied by associated shear and energy dissipation, consider a flat plate with a sharp leading edge and of length x slicing at velocity V directly across an intermaterial area in a fluid of kinematic viscosity ν. To make the picture more realistic assume that the area is associated with a striation of thickness Δx_0.

Immediately, one is faced with boundary layer assumptions which do not serve well the Lagrangian picture of the striation in the boundary layer and in the wake. A laminar wake, fully described, would include a closed recirculation inside the real displacement thickness. Here we shall assume the wake of a plate ending at x is soon cut at right angles by another plate. A simplified boundary layer is used to obtain an approximate picture of the striations in the wake; namely $v_x/V = y'\sqrt{x'}$ for $y' \leq \delta' = 3\sqrt{x'}$ and $v_x/V = 1$ for $y' \geq 3\sqrt{x'}$. The resulting boundary layer is pictured in Fig. 5 along with a striation $\Delta x_0'$ at the initial moment when one side is cut by the

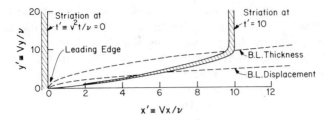

Figure 5   Striation folding and thinning in a boundary layer.

sharp edge and later as a $\Delta y'$ at time $t' = x' = 10$. Definitions of the dimensionless variables are given in the figure. Evolution of the striation is followed as the trajectory of fluid particles in the original intermaterial surfaces.

At $x' \geqslant 6$, the approximate flow field will give a limiting relationship,

$$\Delta y' \rightarrow \frac{12}{5} \frac{\Delta x_0'}{(x')^{1/2}} \tag{9}$$

so that $x' \geqslant 6$ before $\Delta y' < \Delta x_0'$ and $x' = 24$ before the shear stretched striation would be cut in half by one cut. The mechanical energy input rate per unit width of blade (both sides) is $(4\rho V^2 \nu / 3 g_c)\sqrt{x'}$, and the intermaterial area increase per unit energy input works out to be $(3/2)\,(1 - (12/5)\sqrt{x'})(g_c / \nu \rho V \sqrt{x'})$.

Consider now the previous description of a single striation formed by passage of a blade through an intermaterial area. At some distance behind the blade where $u(y = 0) \rightarrow 0$ the single striation thickness will be twice that of the displacement thickness $\Delta y_0' = 3\sqrt{x'}$, that is, the boundary layer thickness at the trailing edge x'. When the wake is cut cross-wise, this $3\sqrt{x'}$ is $\Delta x_0'$ and $\Delta y_1' = (12/5)3(x')^0$, for thinner striations bounding a new wake striation of thickness $3\sqrt{x'} - 2(12/5)3(x')^0 = 3\sqrt{x'}(1 - 2(12/5)3(x')^0)$. n-cuts give a striation thickness of the original wake striation equal to $\Delta y_n' / x' = (12/5)3n/(x')^{n+1/2}$ with similar reductions in larger but successively smaller wake striations that formed in subsequent cuts. It is not difficult to visualize development of a distribution of striation sizes for the whole wake region and striation thickening if the striations are cut at too small an angle subsequently. It should also be noted that alternate striations, wake and previous wake, tend to be of different sizes.

An egg beater, which is a truly remarkable mechanical invention, mixes in this way. If the fluid is not too viscous, an induced flow is pumped through the cutting blades and circulates in the mixing bowl. When the fluid is more viscous, the circulating flow is moved through the blades by rotation of the bowl. The four blades, which cover about one-quarter of the periphery of rotation, interlock in such a way that the wake of one blade is cut crosswise by the following blade on the second rotor. In the circulating flow the wakes in horizontal section form an intermeshed pattern of waves which are more finely cut each time the circulating flow passes through the blades.

In a frame of reference following the blade, flow inside the displacement thicknesses will be in the direction of blade motion and be composed of recirculated flow. A similar type of mixing occurs in the internal boundary layer building up from the rear stagnation point, but the outflow is flared back by the main boundary layer flow outside the displacement thickness. Thus, the Eulerian frame for the macroscopic flow in the blade frame must be considered along with the mixing frame to describe mixing in this case. The flow is continuous only with respect to recirculation in the mixing bowl. Within the flow itself the striations are being stretched in the smallest mixing scales and periodically folded on even larger scales.

2.5. Laminar Mixing in a Spreading Line Vortex (Model of mixing in vortices shed intermittently from a flow separation; mixing action of propeller and "coherent structures" in turbulent mixing)

At some initial time ($t = 0$) assume that a line vortex of strength C appears in the intermaterial area between two portions of fluid which are mechanically alike but subsequently are mixed. Initial conditions are pictured in Fig. 6a. The tangential velocity v_θ of an intermaterial fluid particle is now given by

$$v_\theta = \frac{C}{2\pi r} \left[1 - \exp\left(-\frac{r^2}{4\nu t} \right) \right] \qquad (10)$$

and the position θ of an intermaterial particle beginning and remaining at radius r is given later by

$$\theta' \equiv \frac{1}{C'} \left(\frac{\theta}{2\pi} \right) = t' \left\{ 1 + t' \exp\left(-\frac{1}{t'} \right) \right\} \qquad (11)$$

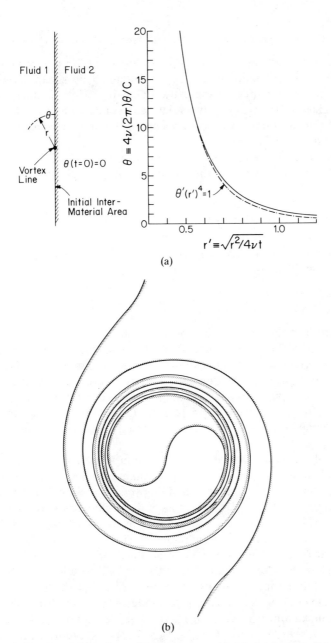

Figure 6 Laminar mixing in a diffusing vortex, (a) Ideal initial condition and equation of double spiral formed. A line vortex cannot be attained, and the spiral at small radii is not realized. An initial vortex is formed with a cylindrical core of finite size. (b) Striations in double spiral of diffusing vortex a short time after formation. The radius of the solid-body core will be of the order of the thickness of the vortex sheet from which the vortex is formed.

where $t' \equiv 4\nu t/t^2$ and $C' \equiv C/4\nu(2\pi)^2$. Since dv_θ/dr is always finite, this is a case of laminar mixing accompanied by shear. However, shear cannot be cited as the cause of mixing. Mathematical model analogies occur in unsteady radial heat conduction, and the combined time and space variable appears in the previous description of boundary layer mixing.

Striations are delineated by a double spiral which is centered on the vortex line. At time t the spirals $\theta = \theta(r)$ are given by

$$\theta' = \frac{1}{(r')^2}\left\{1 + \frac{1}{(r')^2}\exp\left(-(r')^2\right)\right\} \tag{12}$$

where $r' \equiv (r^2/4\nu t)^{1/2}$. At small enough r', that is, at long times, the spiral simplifies to

$$\theta'(r')^4 = 1 \qquad \text{when } r' \leq 0.5 \tag{13}$$

A picture of the spiral is shown in Fig. 6b.

The spiral at $r' \to 0$ is unrealistic because the vortex has to be formed originally as a fluid cylinder of finite size in nearly solid body rotation. Figure 7 is a schematic picture overlay which interprets, from classic photographs[16], mixing in flow downstream of a grid. This train of vortices with increasing age is similar to the train of vortices behind any bluff object immersed in a flow or moving in a fluid.

Initial roll up of this vortex can be approximated by a solid body core of radius $r_c \approx 2(\nu t_c)^{1/2}$, an approximate thickness of the vortex sheet. Here $t_c \approx kL/V$ is the shedding time for successive eddies or alternating eddies, L is the cross stream dimension of the bluff object which causes the vortices and V is the flow velocity (k has a value of order 6). The peripheral velocity $v_\theta(r_c) \approx V$. The double spiral at t_c has achieved only one-half turn and shows a bulbous end on each of the alternate striations. This central core is in nearly solid body rotation, and the bulbs decrease in size very slowly.

To develop further a simplified model which uses Eq. (10), note that a maximum $v_{\theta m}$ occurs at $r_m^2/4\nu t_m = 1.256$ and that $C = 2\pi r_m v_{\theta m}$. If $t_m = t_c \approx kL/V$ and if $v_{\theta m} = V_c \approx V$, C is predicted as $C' \approx (1.57/4\pi)\sqrt{kLV/\nu}$. The spiral will have bulbous ends at $r < r_c = \sqrt{(1.256k)4\nu L/V}$. At later times $t > t_c$, $r_c' \approx (1.12)\sqrt{t_c/t}$ gives for the spiral picture of Fig. 4b the approximate location of the poorly mixed core; that is, the bulbous ends of the striations lie inside r_c. Thinnest striations occur just outside this core, and the alternate striations increase in thickness as they extend to larger radii. Again, a mechanism exists for creating a distribution of striation sizes when they are averaged over the volume of the vortex of mixed flow. However, one can make an important observation about local stoichiometry in that alternate striations of nearly the same size are contiguous at striation scales.

Since Eq. (13) holds when $t \gtrsim 4t_c$, it can be used without much further loss of precision to specify the striation picture at $r \gtrsim r_c$ and when $t \gtrsim t_c$. This further simplification gives an approximate inter-material area generated per unit energy cost of the vortex. The energy per unit length of the model vortex formed intermittently in time $t_c \approx kL/V$ will be $(C_D/2)(pV^2/2g_c)kL^2$ where the drag coefficient C_D is of order unity when LV/ν is of order greater than unity. At time $t > t_c$ the intermaterial area per unit length is $(1.57k)/(1.12)^3)(8\pi L/3)(t/t_c)^2$. Comparison of this value with that for boundary layer mixing given in the previous section suggest that (1)

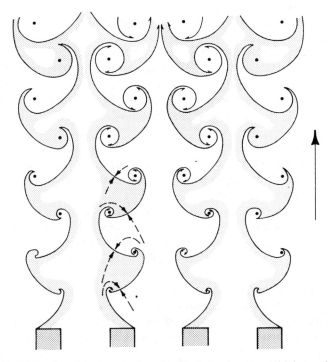

Figure 7   Mixing flow downstream of a grid. The train of vortices with increasing age is similar to the train of vortices behind any bluff object immersed in a flow or moving in a fluid.

on the basis of power input for usual configurations and conditions vortex shedding is desirable, and (2) mixing is dependent on detailed flow geometry as well as on power input.

Mixing has been pictured above as occurring in trailing eddies when L is moved flatwise across the intermaterial area or as depicted in Fig. 7. In Fig. 7 reverse flow in the wake between the two rows of vortices behind a barrier L is seen to be alternately the two different fluids. Movement of L through an already striated fluid would have alined the striations in a tangential direction in the original vortex at t_c. Striations in the backflow of the wake could also be tangentially oriented as they enter the alternate coil of the eddy. Here it would appear that shear has played a role in orienting the striations, but their further mixing or stretching occurs in the eddy itself.

2.6.   Mixing in a Three-Dimensional Laminar Flow between Moving Parallel Plates (Model of mixing in a screw extruder)

To give a quantitative treatment to a mixer related to Fig. 8, imagine that a parabolic feed flow forming an intermaterial area in the orientation shown enters the picture perpendicular to it. The flow now moves in the z-direction between the plates while it is sheared in the y-direction by the moving plate. The intermaterial area could be anchored on a sharp edge ending of an inflow separator. If the upper plate was not moving, the intermaterial area forming in the z-flow would continue to show the same vertical line when projected on the figure. Projection of the intermaterial area in plane $x = 0$ would show an ever increasing parabolic shape.

However, if the upper plate is moving, the intermaterial surface forming is also anchored on the upper plate from $x = 0$ to $x = bGt$. It takes the form of a ballooned sail with again a parabolic-shaped area when projected on plane $x = 0$, a triangular-shaped area when projected on plane $z = 0$, and a parabolic-shaped area when projected on plane $y = 0$. Figure 8 pictures the intermaterial area being developed.

Making a boundary layer-type assumption about relative values of the various shear stresses and assuming a negligible flow adjustment distance in the z-direction, the trajectory of fluid points from $x = 0$, $y = y$ and $z = 0$ forming the intermaterial surface are approximated by

$$v_x = \frac{dx}{dt} = Gy \qquad \text{or } x = Gyt \qquad (14)$$

$$v_z = \frac{dz}{dt} = \frac{3V}{2}\frac{y(b-y)}{(b/2)^2} \quad \text{or} \quad z = \frac{3V}{2}\frac{y(b-y)t}{(b/2)^2} \quad \text{and } y = y \quad (15)$$

where V is the flow averaged velocity in the z-direction and b is the plate spacing. At a given t the maximum z of the area projection on the $y = 0$ plane will be for $y = b/2$ and have the value $3Vt/2$. This maximum will occur at $x = Gbt/2$.

At time t the projected intermaterial area on $x = 0$ is bVt; on $y = 0$ it will be $BVt(Gt)$. At sufficiently large values of Gt the total intermaterial area will be given approximately by the area projection on $y = 0$. Then the ratio of area created by plate shearing to the area created by simple flow contact will be of order Gt. The ratio of final to initial striation size will be $(Gt)^{-1}$; and the effective $\alpha = t^{-1}$, the eventual value of α in Fig. 3 with its imagined striation orientation. The importance of flow geometry in achieving mixing is clearly indicated by the example.

It should also be noted for a continuous flow between plates of finite length in the z-direction that the time of sideways shear is shortest at $y = b/2$ and that most of accumulated flow passes through this central region. The intermaterial area shown in Fig. 8, when viewed as the result of an inflow striation will be much thinner near the moving wall at $y = b/2$ and at larger x. Again, there is cause for the creation of a distribution of striation sizes in the total flow.

In a screw extruder, development of the intermaterial area in the x-direction is soon blocked by the walls of one of the flights, which are causing the z-motion.[17] The stretching area curls into a spiral pattern in z-cross section. Simplified models of the flow field sufficient for estimating the pumping action, will not show details of the mixing action. Here one should interpret coordinate z as the spiral axis between the flights, barrel and screw root surfaces. A parabolic flow, predicted approximately by the pressure gradient of a volumetric flow created by the component of shear in the direction and/or a pressure difference imposed externally, will be observed passing through the $x - y$ plane. Component motion of the barrel surface on the long side of a somewhat rectangular z-cross section imposes a rotation on the flow. The developing intermaterial area in z-cross section curls into a spiral pattern. The center of rotation and spiral will be displaced toward the corner of the $x - y$ cross section into which the plate moves; and the intermaterial area per unit volume in the outlet flow cross section will have to be accumulated and striation values will have to be distributed in terms of the profile of axial flow rates.

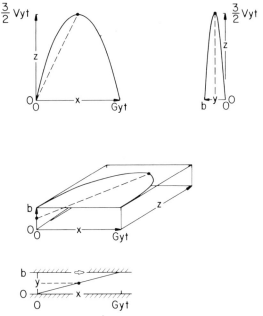

Figure 8   Intermaterial area forming in sideways shear of a parabolic flow. In a screw extruder, development of the intermaterial area in the x-direction is soon blocked by the walls of one of the flights. The stretching area curls into a spiral pattern in z-cross section. The dashed line is the trajectory of the fluid particle shown.

The simplified model shown in Fig. 8 can be continued for an extruder by assuming an ideal flight at which the continued surface reflects in a mirror image across an x-axis at $y = b/2$. However, since particles of the surface make such a cross over in a finite x-length, and in an assymetric pattern the model is becoming less satisfactory for exact predictions even though it shows important qualitative features of the mixing process. In particular, it is noted that $\alpha \rightarrow t^{-1}$ for "axial" flow at each radius from the center of the spiral. The mixing gradient and axial flow time increases with radius such that there is a wide distribution of striations sizes in an outflow valve.

3.   COMMENTARY ON LAMELLAR DESCRIPTIONS

The lamellar description of fluid mechanical mixing is associated with foundations of a theory for structured media. In this example of structuring one considers the smallest macroscopic scales of time and space above those that characterize the molecular actions of classical continua. The complexity of the description is certainly increased, but the degree of approximation can be calculated and is no more than that of boundary layer theory. Consideration of boundary and initial values, of wettability, and of actions of inter-material lines and surfaces—all complications of structure—can lead to a resolution of many paradoxes which have arisen in mixing technology.

At the level of fluid mechanical theory, the time- and boundary-dependent field of vectors which determine $\alpha(t)$ for the Lagrangian flow are a difficult and elegant challenge to further developments of theory. Detailed descriptions of vortex formation and work on coherent structures in turbulent flow are directly applicable to practical problems in mixing. Because they were near the scale of boundary conditions, intermaterial areas considered here tend to be grossly sized. In some cases it is difficult to visualize what $s(t)$ means in these flows. However, in later stages of mixing, when $s(t)$ is small compared to the mixing flows described, the striation surfaces of $s(t)$ are oriented in the same way as the surface depicted and undergo the same stretching process.

The difficulty of understanding fluid mechanical mixing by experimental means has been compromised by including diffusion as a part of mixing and being unable to probe for fluctuations in concentration at scales less than $10 \ \mu m$ where variations in chemical reaction rate occur at scales no larger than $10^{-2} \ \mu m$. If a diffusion controlled, very fast reaction is used to follow the progress of mixing, perhaps by temperature rise, one is often frustrated at some mixing scale by a not-fast-enough reaction. Furthermore, there is a large problem of unravelling mixing, or fluid mechanical mixing, from an averaging process involving a distribution of contact times at reaction temperatures as well as a distribution of boundary values and of $\alpha(t)$ experiences.

When it was first introduced by Mohr, the lamellar description was used in a relatively simple way. It did not seem to offer much more than did statistical descriptions which depend on measurements of concentration fluctuations in a Eulerian frame and which make use of highly developed statistical theories of turbulence[18-20]. The statistical nature of mixing is not displaced by a lamellar description. The lamellar description adds to the statistical description by specifying the actual physical meanings of averages, moments, and distributions of values. The major advantage of a lamellar model is that it is deterministic and suggests, without tedious statistical analysis of data, what causes may effect a desired result in a mixing process.

4. NOTATION

a_v intermaterial area per unit volume

b plate spacing in shear flow mixing

C strength of vortex line in vortex mixing

c_A, c_B concentration of chemical species A in fluid A and B in fluid B. $c_{A0} = c_A(t = 0)$. $c_{B0} = c_B(t = 0)$.

D tank or tube diameter

D_A diffusivity of species A

d tube diameter

d_j orifice diameter in impinging jet mixer

G velocity gradient in shear flow mixing

k first order, homogeneous reaction rate constant; also constant of proportionality for vortex formation.

L characteristic length

R_A volumetric generation rate of species A

r radial position when there is axial symmetry

s striation or material lamina thickness, s_A for fluid A, s_B for fluid B. An average $s = a_v^{-1}$.

t time

V characteristic velocity

V_j jet velocity in impinging jet mixer

v_x, v_y, v_z, v_r components of fluid velocity

w mass flow rate, w_A mass flow rate of fluid A, w_B mass flow rate of fluid B

Y_A dimensionless concentration. $Y_A \equiv c_A/c_{A0}$.

x, y, z space coordinates

Greek Symbols

α velocity gradient in the s-direction in a coordinate system based on the intermaterial area, see Eq. 2.

δ boundary layer thickness

ϵ_A volume fraction of fluid A

ν kinematic viscosity of fluid

ξ dimensionless space coordinate in the direction of s. $\xi \equiv y/s$

ρ fluid density, ρ_A density of fluid A, ρ_B density of fluid B

τ dimensionless warped time. $\tau \equiv \int_0^t [s(t')]^{-2} dt'$

ϕ mixing-reaction parameter for determining mixing or reaction control of an apparent reaction rate, see eq. 6

ψ stream function

REFERENCES

1. Mohr, W. D., Saxton, R. L., and Jepson, C. H. (1957). Mixing in laminar-flow systems, *Ind. Eng. Chem.*, **49**, 1855; (1957). Theory of mixing in the single-screw extruder, *ibid.*, 1857.

2. Ranz, W. E. (1979). Applications of a stretch model to mixing, diffusion, and reaction in laminar and turbulent flows, *AIChE J.*, **25**, 41.

3. Ottino, J. M., Ranz, W. E., and Macosko, C. W. (1979). A lamellar model for analysis of liquid-liquid mixing, *Chem. Eng. Sci.*, **34**, 877.

4. Ottino, J. M. (1980). Lamellar mixing models for structured chemical reactions and their relationship to statistical models; macro- and micro-mixing and the problem of averages, *Chem. Eng. Sci.*, **35**, 1377.

5. Ottino, J. M., Ranz, W. E., and Macosko, C. W. (1981). A framework for description of mechanical mixing of fluids, *AIChE J.*, **27**, 565.

6. Batchelor, G. K. (1967). *An Introduction to Fluid Dynamics*, chap. 2, p. 82, *Kinematics of the Flow Field*, Cambridge Univ. Press.

7. Fisher, D. A. (1968). A model for fast reactions in turbulently mixed liquids, M.S. Thesis, University of Minnesota, Minneapolis; (1974). Development and application of a model for fast reactions in turbulently mixed liquids, Ph.D. Thesis, University of Minnesota, Minneapolis.

8. Torrest, R. S., and Ranz, W. E. (1969). An improved conductivity system for measurement of turbulent concentration fluctuations, *I.E.C. Fund.*, **8**, 810; (1970). Concentration fluctuations and chemical conversion associated with mixing in some turbulent flows, *AIChE J.*, **16**, 930.

9. Vassilatos, B., and Toor, H. L. (1965). Second-order chemical reactions in a monhomogeneous turbulent fluid, *AIChE J.*, **11**, 666; Mao, K. W., and Toor, H. L. (1970). A diffusion model for reactions with turbulent mixing, *AIChE J.*, **16**, 49.

10. Rys, P. (1976). Disguised chemical selectivities, *Accounts Chem. Res.*, **9**(10), 345; (1977). *Agnew. Chem.*, Int. Ed. Engl., **16**, 807.

11. Bourne, J. R., Rys, P., and Suter, K. (1977). Mixing effects in the bromination of resorcin, *Chem. Eng. Sci.*, **32**, 711; Bourne, J. R., and Kozicki, F. (1977). Mixing effect during the bromination of 1,3,5-trimethoxybenzene, *ibid.* **32**, 1538.

12. Kwon, H. S. (1976). A description of mixing of fluids based on intermaterial area per unit volume, M.S. Thesis, University of Minnesota, Minneapolis.

13. Danckwerts, P. V. (1952). The definition and measurement of some characteristics of mixtures, *Appl. Sci. Res.*, **A3**, 279.

14. Ou, Jane-Jane, and Ranz, W. E. (1979). Mixing and chemical reactions: part 1, A contrast between past and slow reactions: part 2, Chemical selectivities; part 3, Thermal effects, Papers submitted for publication based on Ou, Jane-Jane, Mixing and chemical reactions, Ph.D. Thesis, University of Minnesota, Minneapolis.

15. Goldsmith, H. L., and Mason, S. G. (1976). *Rheology*, Fig. 52—Time reversed flow: an "unmixing" demonstration, in *The Microrheology of Dispersions*, Vol. 5, pp. 85–250. *Rheology* (Ed. F. R. Eirich), Academic Press, New York.

16. Prandtl, L. (1938). Proc. 5th Int. Congr. App. Mech., Cambridge, Mass., p. 340; also reproduced in Hinze, J. O. (1975). *Turbulence*, 2nd Edn, McGraw-Hill, New York.
17. Tadmor, Z., and Gogos, C. G. (1979). Section 10.3 the screw pump, and section 11.10 screw extruder in chapter 11 mixing, *Principles of Polymer Processing*, Wiley, New York.
18. Corrsin, S. (1957). Simple theory of an idealized turbulent mixer, *AIChE J.*, **3**, 329; (1964). The isotropic turbulent mixer: part II, arbitrary Schmidt number, *Ibid.*, **10**, 870.
19. Murthy, S. N. B., Ed. (1975). *Turbulent Mixing in Non-reactive and Reactive Flows*, Plenum Press, New York.
20. Patterson, G. K. (1981). Turbulent mixing with chemical reaction, Proc. 2nd World Congress of Chemical Engineering, Vol. IV, Unit Operations, Montreal Canada, Oct. 4–9.

Chapter 2

Fundamentals of Turbulent Mixing and Kinetics

ROBERT S. BRODKEY

Department of Chemical Engineering, The Ohio State University, Columbus, OH 43210, USA

1. INTRODUCTION

Mixing in one form or another has been used as a scapegoat in process analysis, especially when chemical reaction is involved. Whether the blame is justified or not can only be resolved by an understanding of the mechanism of mixing and its interaction with chemical kinetics. This review will attempt to set the stage for the two reviews that follow. To do so requires that the source of the mixing be described. To be considered here is turbulent mixing; laminar mixing, as a combination of folding and molecular diffusion, will not be treated as it was considered in some detail in the first review. The material in this review is a composite from several recent sources[1-4] and should be pictured as an abbreviated introduction to the field of turbulent motion, mixing and kinetics. Without the references cited above it is incomplete. The purpose in bringing the material together is to provide an introduction to those unfamiliar with the field.

What needs to be made clear in this review is what we mean by mixing, dispersion, blending, diffusion, convection, etc. How can these terms be quantized and what means can be used for their measurement? What are the experimental methods involved in the measurements and what are some of the results? What simple theories of mixing have been proposed and what are some of the consequences? And finally, what are the extensions for chemical reactions and some of the results for such systems?

The term mixing is a loose one and encompasses nearly as many definitions as there are workers in the field. This review will deal with local or fine scale mixing, that is mixing that we would like to have when we are interested in encouraging chemical reaction between two species. We would like to know how the molecular diffusion of the individual species and the violent turbulent motions interact to bring the molecules together for reaction to occur. We must remember that molecular diffusion is an essential part of the sequence of events necessary for chemical reaction between species. Since molecular dimensions are orders of magnitude smaller than the smallest of turbulent scales, molecular diffusion is a necessary step for bringing two species together for chemical reaction.

Turbulence is a difficult subject; it is still more difficult to combine the effects of turbulence and molecular diffusion and to incorporate kinetics is a further complication which makes the analysis most difficult. Thus, it is not surprising that approaches to this subject involve idealized models. What we would like to be able to do is pictured in Fig. 1. From a known geometry we would like to be able

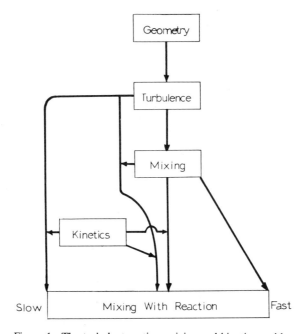

Figure 1   The turbulent motion, mixing and kinetics problem.

to predict or, if necessary, measure the parameters of turbulence. From this we want to obtain the mixing. Then, with incorporation of kinetics, we want to predict the full range of mixing with chemical reaction from the slow self-mixing (back-mixing) to the fast reaction limit.

1.1. The Mixing Process, An Overview

We are interested in what is called local mixing or fine scale mixing. It will help to put this type of mixing into the context of mixing in general. Our idea of mixing depends, to a great extent, on our definition of the term mixture. We shall use mixing to mean any blending into one mass, and mixture to mean a complex of two or more ingredients which do not bear a fixed proportion to one another and which, however thoroughly co-mingled, are conceived as retaining a separate existence. In any specific case, our mixture will depend on the scale of our view. If our scale of view is very large, then even a very coarse mixture may be all we want, and molecular diffusion means little to us. Here, mixing, blending and dispersion are all equivalent expressions. But if we want a mixture on a finer scale, then dispersion alone will not do; we must have fine scale mixing by the smallest of eddies in the turbulence in conjunction with molecular processes. We will have to look at each in even more detail if we are to obtain a real feel for the mixing process.

There is a paradoxical point that should be mentioned first. The mixing of two individual species (including molecular diffusion effects as we must) can give rise to a mixture of the species since the molecules of the two species can always be distinguished. The same is not true for heat or for momentum since the final blend of these are completely homogeneous, in that, at the end of mixing, we cannot distinguish a hot element of fluid or an element with a high velocity from one that had been cold or had a different velocity. We will not let this bother us since the basic processes are quite similar, especially for the scalar quantities of heat and mass, and all are often referred to as mixing processes.

1.2. Terminology in Mixing

Let us proceed by briefly discussing each of the processes that are usually considered as one form or another of diffusion, dispersion and mixing. The word *diffusion* means the act of spreading out; it connotes nothing of the mechanism providing the spread. However, the unmodified word usually implies diffusion by molecular means. Since other processes can give rise to diffusion and, indeed, are called *diffusional processes*, the term *molecular diffusion* is better used to signify diffusion caused by relative molecular motion. The term dispersion is usually not used for the molecular process but otherwise is interchangeable with diffusion and will be used as such in this discussion. In turbulent flow, there is bulk motion of large groups of molecules. These groups are called *eddies* and give rise to the material transport called *eddy diffusion* or *eddy dispersion*. Non-molecular and non-eddy diffusion processes can be grouped into a class which we will call *bulk diffusion* or *dispersion*. In each case, there is some bulk motion or convection giving rise to dispersion which is superimposed on either molecular or eddy diffusion or both. For example, in turbulence, the problem is complicated by three superimposed diffusional processes. Molecular diffusion is always occurring and may not be always neglected. Superimposed is the gross eddy motion causing the eddy diffusion. Finally, it is possible to have other types of bulk diffusion occurring simultaneously. To explain a little further, bulk diffusion is usually pictured as a result of specific convection mechanisms or large scale motions that are causing the dispersion. When the mechanism is not clear, it is usually considered a statistical problem and included as a part of the eddy dispersional process. For example, circulation of fluid in a tank is convection that has often been considered as a large scale turbulent motion rather than a bulk dispersion process. More recently tank circulation has been considered as convection. Sometimes, though, the diffusional concept has been carried to the extreme as when one defines a negative eddy viscosity. It is the combination of molecular, eddy and bulk or convection effects that is best described by the term *mixing*.

1.3.   Interactions of the Mixing Processes

Molecular diffusion is a product of relative molecular motion. In any fluid where there are two kinds of molecules, if we wait long enough, the molecules will intermingle and form a uniform mixture on a submicroscopic scale (by submicroscopic, we mean larger than molecular, but less than visual by the best microscope). This view is consistent with the definition of a mixture for we know that if we were to use a molecular scale, we would still observe individual molecules of the two kinds, and these would always retain their separate identities. The ultimate in any mixing process would be this submicroscopic homogeneity where molecules are uniformly distributed over the field; however, the molecular diffusion process alone is generally not fast enough for present-day processing needs. In some systems, molecular diffusion is so slow as to be completely negligible in any reasonably finite time; high molecular weight polymer processing is a good example of this state.

If turbulence can be generated, then eddy-diffusion effects can be used to aid the mixing process. For some materials, the generation of turbulence would be too expensive because of high viscosity, and, in others, it might be impossible because of product deterioration under the high energy inputs required. In those instances mechanical means of promoting mixing are usually employed.

Each of the bulk-diffusion or dispersion phenomena tends to reduce the scale by spreading a contaminant over a wider area. The molecular diffusion is enhanced because of the larger area. It is important to note that if the molecular diffusion is rapid enough, the system may be almost submicroscopically mixed by the time the bulk diffusion has spread the contaminant over the field. With low rates of molecular diffusion, this will not be true.

2. THE MECHANISM OF MIXING

2.1. General Description

The type of mixing of concern is the mixing between two or more streams. We have component A in one stream and component B in the other. We want to mix these into a homogeneous system where molecules of A and molecules of B can come into contact. The smallest scale of turbulence is large compared to molecular size, and the smallest eddy contains millions of molecules. Turbulence can only reduce the size to that of its smallest eddy, and thus we would still have many A molecules in one region and many B molecules in an adjacent region. One must have molecular diffusion in order to bring the A and B molecules together to react. This is the interaction between turbulent motion and molecular diffusion. This mixing should be contrasted to another type often referred to as a back-mixing, or better as self-mixing. Self-mixing is the mixing of a homogeneous fluid entering a mixing vessel. Here, material recently introduced is mixed with material that has been in the vessel for a longer period of time. This is a mixing in time and not a mixing between components in the same sense as described above.

As it is usually formulated, this mixing does not involve molecular diffusion, since it is usually assumed that the local molecular mixing is instantaneous when the old element comes in contact with a newer one.

Next, let us consider a model of the details of the process as the system proceeds from an unmixed state to a mixed one. Let us take two streams that are in some way separately identifiable. Figure 2 illustrates the initial process of *dispersion*, which is defined as the act of breaking apart and causing to go different ways. Turbulent dispersion does not necessarily involve molecular diffusion processes at all; in fact, the molecular diffusivity does not even enter the dispersion problem as described by Taylor[5]; its effect was added by others later. A uniform dispersion or mixture can be obtained. One can continue to reduce the size of the blobs until they can be reduced no further by the action of the turbulence which does have a size below which it is ineffective. At the same time, molecular diffusion is usually acting as shown in Fig. 3. The final, very fine scale uniform mixture must come from the molecular mixing produced by diffusion. Note that in Fig. 3, the dispersion process of Fig. 2 has been completed to a level as given by L_s. Three of these levels are shown.

There are two criteria for mixing which, for now, we can treat qualitatively and later define more quantitatively. The first is the *scale of segregation* (L_s): a measure of the size of unmixed clumps of pure components. This is a measure of some average size. As the clumps are pulled and contorted, the scale of mixing is reduced; this would be going from left to right along the top of Fig. 3. The second criterion is the *intensity of segregation* (I_s), which describes the effect of molecular diffusion on the mixing process. It is a measure of the difference in concentration between neighboring clumps of fluid. The intensity, for each value of the scale, is illustrated by the columns in Fig. 3.

To summarize, the turbulent process can be used to break up fluid elements to some limiting point; however, because of the macros-

Figure 2   The dispersion process (reduction of L_s).

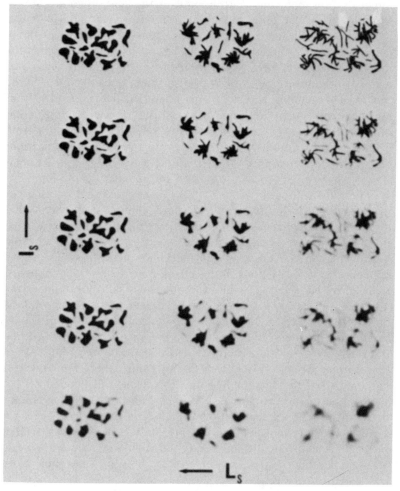

Figure 3   The mixing process (reduction of I_s and L_s).

copic nature of turbulence, one would not expect the ultimate level of breakup (scale) to be anywhere near molecular size. Since energy is required for this reduction in scale, the limiting scale should be associated with the smallest of the energy containing eddies, that is, above the energy dissipation range. This size will be large when compared with molecular dimensions. No matter how far we reduce the scale, we still have pure components. Depending on the size observed, any one of these levels in scale might be considered mixed; however, from a view of submicroscopic homogeneity, where molecules are uniformly distributed over the field, none are mixed. Without molecular diffusion, ultimate mixing cannot be obtained.

Molecular diffusion allows the movement of the different molecules across the boundaries of the elements, thus reducing the difference between elements. This reduction in intensity will occur with or without turbulence; however, turbulence can help speed the process by breaking the fluid into many small clumps, and thus allowing more area for molecular diffusion. When diffusion has reduced the intensity of segregation to zero, the system is mixed. The molecules are distributed uniformly over the field. Various degrees of this combined process are shown in Fig. 3. In systems where reaction is to occur, the need for submicroscopic mixing is apparent, for without it, the only chemical reaction that would occur would be on the surface of the fluid clump. Danckwerts[6] has discussed the importance of this degree of mixing of two reactants; the intensity of segregation must be reduced rapidly so as to avoid local spots of concentrated reactant and the usually associated undersirable side reactions. In jet mixing, the scale of segregation is reduced by eddy motion while the molecular diffusion reduces the intensity. In a jet, if a solid product occurs, its particle size will be a function of the rate of reduction of segregation. The same would be true in the quenching of a jet of a hot gas reaction mixture; the freezing of the reaction products will depend on the reduction of segregation. Another example is the jet flame, where the oxygen is obtained from the surrounding air. The flame will depend on the segregation of the two gases. In laminar flames, the mixing is poor because the scale of segregation is high. The flame occurs along a surface and is controlled by the molecular diffusion across that surface. In a turbulent flame, eddy diffusion will reduce the scale and provide more area for molecular diffusion and thus more contacts for burning.

2.2. Quantitative Description of Mixing Criteria

Two aspects of mixing have been mentioned: the degree to which the material has been spread out by the turbulent action (scale of segregation) and the approach to uniformity by the action of molecular diffusion (intensity of segregation). Danckwerts[7] has provided definitions for these in terms of the statistical variables that can be measured in the mixing field. The criteria cannot be applied where gross segregation occurs. This restriction implies that the concentrations are grossly uniform over the field, but not necessarily uniformly mixed on anything but the largest scales.

The *scale of segregation* (L_s) is an average over relatively large distances, and thus is a measure of the large-scale breakup process, but not of the small-scale diffusional process. It is defined as follows:

$$L_s = \int g_s(r)\, dr \tag{1}$$

where $g_s(r)$ is the Eulerian concentration correlation given by

$$g_s(r) = \frac{\overline{C'_A(\mathbf{x})C'_A(\mathbf{x}+\mathbf{r})}}{\overline{C'^2_A}} \tag{2}$$

and $C'_A = C - \bar{C}_A$. The concentration, C_A, is expressed as a concentration fraction where 1 is the maximum value. In liquids with very slow molecular diffusion, the scale would decrease to some limiting value dependent upon the distribution of globs as caused by the turbulent field. This could happen before the fine scale mixing by molecular diffusion progresses very far. In gases, where molecular diffusion is rapid, the scale may not be reduced appreciably before diffusional effects become important.

The *intensity of segregation* (I_s) is defined in terms of the concentration fluctuations, and is measured at a *point* for a long enough time to obtain a true average. It is defined as follows:

$$I_s = \frac{\overline{C'^2_A}}{\bar{C}_{A0}(1 - \bar{C}_{A0})} = -\frac{\overline{C'_A C'_B}}{\bar{C}_{A0}\bar{C}_{B0}} = \frac{\overline{C'^2_A}}{\overline{C'^2_{A0}}} \tag{3}$$

Here the subscript 0 refers to the initial value. The form with C'_A and C'_B is the intensity of segregation when two components in two separate streams are being mixed. The intensity of segregation is unity for complete segregation (i.e., no mixing) and drops to zero when the mixture is uniform (the fluctuations are zero).

If there were no diffusion, and only the smallest possible eddies were present, the value of I_s would still be unity; thus the intensity of segregation as defined is· a good measure of the diffusional process. The intensity of segregation is defined as a function of time-averaged variables at a point. This of course assumes that such an average can be obtained; that is, that the system is at steady state or is changing slowly when compared with the time necessary to obtain the average. For a complete definition of a given system, one would have to specify the variation of I_s over the entire volume. As a simple example, consider plug flow, in which two fluids are to be mixed. It will be assumed that each fluid is initially distributed uniformly across the pipe cross section on a macroscopic scale under the condition of complete segregation ($I_s = 1$). As the fluid moves down the tube in plug flow, mixing will occur as a result of the turbulent field and diffusion, and the value of I_s will decrease to zero in the limit of molecular uniformity. Actually, I_s must be measured over some small but finite volume. If this volume is too small, submicroscopic variations are detected (statistical fluctuations in the number of molecules present), and if the volume is too large, the measurement becomes insensitive and approaches the average value of the system. For many problems (such as nonideal mixers used for reactors), a detailed study of the variation of I_s over the entire reactor is not desirable, and some space average of the entire system is used. Finally, one can contrast the measurement of I_s with the more conventional "'mixing times'", which are a direct function of the means of measurement and thus one cannot attach absolute meaning to them.

3. EXPERIMENTAL MEASUREMENTS

Experimental results on mixing are limited, but enough results on a variety of geometries are available to allow some understanding of the mixing process and the evaluation, in part, of theoretical ap- proaches. Mixing and the turbulence in the core region of a pipe (with a centrally located jet injector)[8], in the impeller stream of a continuous flow-stirred tank[9], and in a multijet tubular reactor configuration[1] have all been measured. Only the results from the tubular reactor will be considered in this brief review.

The objective of the reactor studies was to establish the concen- tration and velocity fields without chemical reaction in reactors similar to those used by Vassilatos and Toor[10] and Mao and Toor[11] to study chemical reactions under turbulent flow conditions. Their studies and that of McKelvey et al.[1] combine to furnish the only complete set of data on turbulent mixing with chemical reaction. The results of the research will be reviewed later after some introductory comments on the theory and mixing are presented. The reactor configuration had a feed module that separated the two feed streams into many alternating small streams so as to provide a grossly mixed initial condition. The mixing was completed down- stream in a pipe that followed the feed module. The reactor and a light probe used for the concentration detection[12] is shown in Fig. 4.

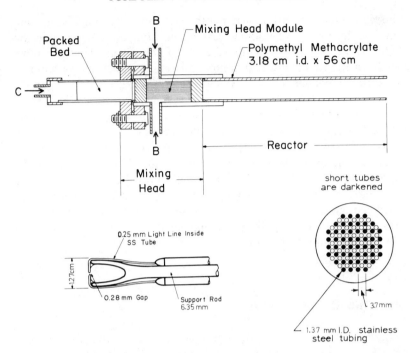

Figure 4   Mixing reactor, feed module, and light probe.

The flow in the reactor was quite complicated[1], especially in the region immediately downstream from the injector plane. The measurements in this region depended on the location of the hot-wire anemometer used to measure velocities (centered on one of the injector tubes or located between two injector tubes). Beyond the jet coalescence plane the results merged. The mean axial velocity centered on a jet decayed rapidly with distance from the feed module. The velocity measured between two adjacent injector tubes rose rapidly and merged at the coalescence point with that measured on a jet center. After coalescence of the jets, the velocities decayed slowly over the range where reaction would occur. The velocity fluctuations initially rose from a low value associated with the turbulence in the small tube or that associated with the region between two jets. The results merged at the coalescence plane and continued to decrease as expected, but then unexpectedly increased. This nonideal flow is attributed to the formation of a large vortex or separation along the wall near the entrance of the reactor. Results on correlations and spectra were used to compute the characteristic turbulence parameters: velocity macroscale and microscale, low wave number cutoff, and kinetic energy dissipation rate. These are essential for calculating the mixing from a knowledge of the turbulent field. Empirical estimates for the parameters require knowledge of the axial turbulent velocity intensity and a characteristic dimension of the flow system; unfortunately, we do not know which dimension is characteristic for the reactor, so that the estimates made were of uncertain usefulness.

One-dimensional scalar spectra were measured, but much of the spectral energy lay beyond the cutoff frequency of the light probe and could not be detected[12]. Far enough downstream from the feed module a (−1) power law region was observed. The existence of such a region in mixing at large Schmidt numbers has been suggested by Batchelor[13]. Both the concentration and velocity fluctuations were found to be normally distributed.

4. THEORY OF MIXING

There have been several simplified approaches to the prediction of mixing. These have been valuable because they can provide some insight into the relationship between turbulence and mixing and can suggest to us some of the controlling variables that should be considered in commercial mixing operations. Under somewhat idealized conditions, the theories allow us to estimate the time required for a specific degree of mixing. Under such conditions, one can even, at times, estimate the extent of chemical reaction that will occur in homogeneous reactions involving two separate feeds and in which the degree of turbulence is a contributing factor. In this section, the details of the theory will not be given, but rather we shall emphasize the results and the implications that these have for mixing operations.

The *dispersion* of a contaminant by turbulent motion is of fundamental importance in many problems. To illustrate, the oceanographer would like to be able to determine beforehand the dispersion of wastes discharged into the sea; the degree of air pollution and the proper design of stacks depend on the ability of the turbulent wind to disperse smoke; and the time necessary for blending will depend on one's ability to disperse the material to be mixed. Such dispersions are often called *turbulent or eddy diffusion*; the analysis of such problems for the most part is concerned with the case of no superimposed molecular diffusion, and, in effect, can be considered as the motion of marked fluid particles or elements.

A Lagrangian view can be used to gain some insight into the mechanism. Consider individual elements that leave a fixed point in space: first, for the case of no mean motion, the various elements will be carried from the source by the turbulent eddies. Those caught in a large eddy (with generally large motion) would be expected to be carried further than those that are initially a part of a small eddy. Thus, at any instant in time, there will be a distribution of elements about the point source. This may be easier to visualize by superimposing a uniform mean velocity on the turbulent field. Each element or particle leaving the point source would be expected to deviate from the linear path in a random manner depending on the local nature of the turbulence. The rms (root-mean-squared) deviation for the particles would be observed as a continued divergence, spread, or dispersion as the particles are carried downstream from the point source by the uniform mean velocity. This is an eddy motion and can occur in the absence of molecular diffusion. In this discussion, the distance from the point source divided by the uniform mean velocity has been used to replace time in the first illustration. It was Taylor[5] who considered the diffusion of infinitesimal fluid particles from a point source in a homogeneous isotropic field with no superimposed molecular diffusion. The theory was restricted to eddy diffusion from a source in a static field or from a source moving with a uniform velocity field.

What we shall call the *turbulent mixing problem* is more complicated. We are interested not only in the spread of the material to be moved, but in addition, how it becomes essentially homogeneous with the surrounding fluid. For this we need molecular diffusion. Because of the nature of the problem, little is known about the actual physical mechanism. Consequently, an approach similar to statistical turbulence is used; that is, the problem is formulated rigorously in terms of statistical averages without reference to any specific model. There are two important aspects of the problem. First, experimental information interpreted in terms of the theory may provide some insight into the actual mechanistic contribution of turbulence and of molecular diffusion to mixing; and second, with reasonable approximations for the mixing spectrum and boundary conditions, the theory can be used to predict the time of mixing under specific mixing conditions. Admittedly, we have not progressed in either direction as far as we would like; however, as will be seen, progress has been made.

The statistical theory of turbulent mixing has been developed parallel to the turbulent motion problem. The basic equation for turbulent mixing is that of mass (or heat) conservation which is the counterpart of the nonlinear Navier–Stokes equation for turbulent motion. It is obvious that the treatment of mass (or heat), which we shall call *scalar quantity*, is much simpler than that of turbulent velocity which is a vector quantity. However, the problem of turbulent mixing of a scalar quantity enjoys all the difficulties that turbulent motion does because of the nonlinearity of the governing physical equations. The velocity fluctuations are always introduced as a part of the unknown functions in the course of describing turbulent mixing in a turbulent field. Therefore, the evaluation of the various functions for turbulent motion, such as correlation and spectrum functions, must be available before one attempts to solve the turbulent mixing problem. The closure of the infinite set of moment equations in mixing is analogous to that of the motion problem. If the analysis of turbulent mixing in an isotropic homogeneous turbulent field (the most idealized physical reality) can be applied to real shear problems, one can arrive at some sort of approximation to the mixing problem. If local isotropic turbulent conditions exist, then the approximation may be good; however, few experiments to indicate the degree of approach to local isotropy have been reported. For complex shear flows, one would not expect the simplification of isotropic conditions to apply on the large scale; thus, more complicated modelling must be used. This involves using the full set of motion and scalar equations plus assumptions about the many unknown terms to allow closure and a solution with a given geometry and boundary conditions. The fact that the analysis is soundly based on the Navier–Stokes equations, continuity of a given species and logically dimensional arguments, sets the stage for a reasonable solution. The mathematics are often complex and the computer solutions are lenghty. However, little is learned about the mechanism of mixing from this approach.

The basic equations involved should be briefly reviewed for further use. The Navier–Stokes equation for an incompressible fluid with constant properties and no external forces is

$$\frac{\partial \mathbf{U}}{\partial t} + (\mathbf{U} \cdot \nabla)\mathbf{U} = -\frac{1}{\rho}\nabla p + \nu\nabla^2\mathbf{U} \qquad (4)$$

The equation for continuity (overall mass balance is)

$$(\nabla \cdot \mathbf{U}) = 0 \qquad (5)$$

and the relation for the mass balance on an individual species is

$$\frac{\partial C_A}{\partial t} + (\mathbf{U} \cdot \nabla)C_A = R_A + D\nabla^2 C_A \qquad (6)$$

where R_A is the generation of the species "A" by chemical reaction. Each of these equations can be time averaged to give

$$\frac{\partial \bar{\mathbf{U}}}{\partial t} + (\bar{\mathbf{U}} \cdot \nabla)\bar{\mathbf{U}} = -\frac{1}{\rho}\nabla \bar{p} + \nu\nabla^2\bar{\mathbf{U}} - (\nabla \cdot \overline{\mathbf{U}'\mathbf{U}'}) \qquad (7)$$

$$(\nabla \cdot \bar{\mathbf{U}}) = 0 \qquad (8)$$

and

$$\frac{\partial \bar{C}_A}{\partial t} + (\bar{\mathbf{U}} \cdot \nabla)\bar{C}_A = \bar{R}_A + D\nabla^2\bar{C}_A - (\nabla \cdot \overline{\mathbf{U}'C_A'}) \qquad (9)$$

The added term in Eq. (7) is the Reynolds stress and is the source of the closure problem previously cited. It is this specific term and the corresponding term in Eq. (9) that require assumptions or further equations to be developed. Additional equations that can be developed are obtained from the foregoing Eqs. (4)–(9) by complex manipulations. The resulting equations are expressions for $\overline{\mathbf{U}'\mathbf{U}}$, $\overline{\mathbf{U}'C_A'}$, $\overline{C_A'^2}$, etc., and certain sums of terms. These additional equations obviously add considerable complexity to any modelling endeavor. The reaction rate term $\bar{R}_A$, in Eq. (9), is very important in considering turbulent mixing and chemical reaction and will be considered in more detail in a subsequent section.

Before embarking on a discussion of some of the theoretical results, we would be remiss if we did not mention the newer research on coherent structures in turbulence and the implications to mixing. A great deal of work is being done on understanding the fundamental physical mechanism of turbulent shear flows, in particular, studies about large scale motions in the flow away from the boundaries and small scale motion in the vicinity of the wall or shear zone. This information to date has not progressed far enough to allow the formulation of a reliable model for application to the mixing process, but the hope exists that this new approach will eventually allow a more general solution to the mixing problem. One can visualize that if a reasonably complete picture of the turbulent motions as a sequence of coherent events can be established, then the effect of these motions on the transport of a scalar quantity could be established.

The theory of Beek and Miller[14] based on a suggestion by Corrsin[15] was one of the first analyses of the scalar mixing problem. The analysis was based on the idea of an eddy-diffusivity type of transfer function that paralleled an earlier suggestion by Heisenberg for the velocity field. Beek and Miller integrated the basic equation under certain idealized assumptions to obtain the intensity of segregation as a function of distance down an idealized pipe line in which there was no velocity distribution (plug flow). The results are quite logical even though they may not be correct in magnitude because of the assumed spectrum. In Fig. 5, mixing of gas in a pipe line is considered. For the mixing of gas in a pipe line, increasing the number of injection points is associated with a reduction in scale which in turn provides a larger area for rapid diffusion. The results for liquid systems (Fig. 6) show a modest dependency on the number of injectors. Here, the diffusivity is so low that the turbulence has enough time to distribute the clumps of material into their smallest eddy sizes before much diffusion has occurred. Thus, one would expect that liquid mixing would be less dependent on initial conditions or number of injectors. The time of mixing for both the liquid and gas system is highly dependent upon the Reynolds number. The decrease in mixing time for the gas just compensates for the increase in velocity associated with the higher Reynolds number. For liquids, the effect of the increase in turbulence more than off-sets the increase in velocity, thus resulting in a net decrease

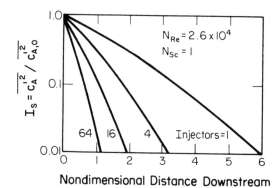

Nondimensional Distance Downstream

Figure 5   Theoretical estimate of gas mixing by Beek and Miller.

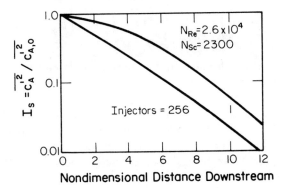

Figure 6   Theoretical estimate of liquid mixing by Beek and Miller.

in the length of pipe required. This latter result is open to question from similarity considerations.

It has often been suggested that the Reynolds number be used for scaleup of pipeline mixers. For constant Reynolds number to obtain the same degree of mixing, one must also increase the length in proportion to the increase in d_0. This is, in effect, a requirement of geometric similarity. Even scaleup on the same velocity will not be sufficient. In all cases, to obtain the same degree of mixing, one must also design for an even greater length than in the smaller test section. Toor and Singh[16] have recently considered scaleup in much more detail. The results are the same for the constant Reynolds number case and are developed in more detail for the variable Reynolds number case.

The conservation of matter equation (the equation for $\overline{C_A'^2}$) can be used to give

$$I_s = e^{-t/\tau} \tag{10}$$

where τ is the time constant of mixing. This has been suggested by Corrsin[17], who has estimated the mixing time constant from spectrum arguments[18].

For gas systems, he obtained

$$\tau = \left(\frac{5}{\pi}\right)^{2/3} \frac{2}{3 - N_{Sc}^2} \left(\frac{L_s^2}{\epsilon}\right)^{1/3} \tag{11}$$

and for liquids

$$\tau = \frac{1}{2}\left[3\left(\frac{5}{\pi}\right)^{2/3}\left(\frac{L_s^2}{\epsilon}\right)^{1/3} + \left(\frac{\nu}{\epsilon}\right)^{1/2}\ln N_{Sc}\right] \tag{12}$$

where N_{Sc} is the Schmidt number and ϵ is the turbulent kinetic energy dissipation.

In many cases one is only interested in scaleup of mixing units. The following is a brief summary of Corrsin's results for this practical problem. In the equations of the time constant of mixing [Eqs. (11) and (12)], the most important term is L_s^2/ϵ. The turbulent energy dissipation can be represented by the power input per unit mass and an efficiency:

$$\epsilon = \eta P/M \tag{13}$$

where η is the efficiency of turbulent production, P is the power and M is the fluid mass. Combining gives the scaleup as

$$\eta' P'/M'L_s'^2 = \eta P/ML_s^2 \tag{14}$$

The mass scales with the geometry or L^3 and the mixing scales as L_S. If the efficiency varies with the scaleup, then the result for the power is

$$P' = (\eta L'^3 L_s'^2/\eta' L^3 L_s^2)P \tag{15}$$

If efficiency is constant, the scale-up is

$$P' = \left(\frac{L'}{L}\right)^3\left(\frac{L_s'}{L_s}\right)^2 P. \tag{16}$$

Finally, if both the mass and the scalar scale-up are the same, the scale-up would be

$$P' = \left(\frac{L'}{L}\right)^5 P \tag{17}$$

While L depends on geometry, L_s would be expected to depend more on the nature of the injection of the material to be mixed. The fifth power can be derived also from pure blending relations (i.e., the number of tank turn-overs held constant on the scale-up); however, the empirical value is nearer to the fourth power than the fifth power. From the equations above, one would expect this to be between the third and fifth power.

Beek and Miller's approach (Figs. 5 and 6) and Corrsin's analysis based on Eqs. (10)–(12) are simple, useful results for mixing analysis as will be shown in the next section where comparisions between experiments and theory will be made. One alternate fundamental approach would be to use Eq. (6) directly with the instantaneous velocity obtained from the modern concept of coherent structures and not from a solution of Eq. (14). Such an approach was recently taken by Brodkey et al.[19] for the case of mass transfer and mixing from a wall to a fluid at high Schmidt numbers. The model provides a measure of the rms concentration fluctuation (and thus $\overline{C_A'^2}$ as a measure of mixing) and as such deviates from the many models in the literature that are only concerned with the mean mass transfer rate at the wall.

5. EXPERIMENTS AND COMPARISONS TO THEORY

5.1. No Chemical Reaction

Comparisons of experimental data with the theories can be made for both the turbulent and scalar fields for the reactor geometry previously described. The theory for a stationary field is used to estimate the mixing at specific locations. Even though crude, because of uncertainty of the characteristic dimension, order-of-magnitude or better estimates can be made for the decay of the scalar fluctuations. This is the same theory earlier used by Brodkey[2] for pipe flow mixing. The conversion in very rapid reactions is directly related to this decay, and thus, once the mixing is predicted, so is the chemical conversion in the case of very rapid reactions.

The most important single statistic of the scalar field is I_s (Fig. 7). Downstream of the coalescence plane, the decay is well represented by

$$I_s = (1.28 \times 10^{-5})t^{-3/2} \tag{18}$$

The time (in seconds) used in plotting Fig. 7 is the average time required for a fluid element to flow from the head to the position at which the measurements were taken.

Predicted time constants [Eq. (12)] at specific points are further compared to the measurements in Fig. 8. A forward integration in time with knowledge of the variation of the time constant of mixing should allow prediction of the entire decay curve. The data of Nye and Brodkey[20] for the decay along the center line of a pipe flow are

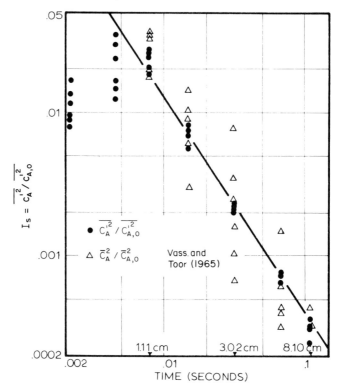

Figure 7   Decay of the concentration fluctuations downstream in a tubular reactor with mixing head injection. Segregation decay $(\overline{C_A'^2}/\overline{C_{A0}'^2})$ is compared to concentration reduction $(\bar{C}_A^2/\bar{C}_{A0}^2)$ for a very fast reaction.

shown for comparison. The faster mixing in the multi-jet reactor is quite apparent.

5.2.   Fast Chemical Reaction

One main purpose is to demonstrate the equivalence of the mixing results in the reaction and the very fast reaction results in the identical geometry as measured by Vassilatos and Toor[10]. The suggested equivalence is a result of the theory of Toor[21], which is a powerful tool for mixing studies. To demonstrate the equivalence from this work, Vassilatos' data, in the form of I_s for very rapid reactions, are shown in Fig. 7 along with the mixing data without reaction. The intensity of segregation is simply calculated from the

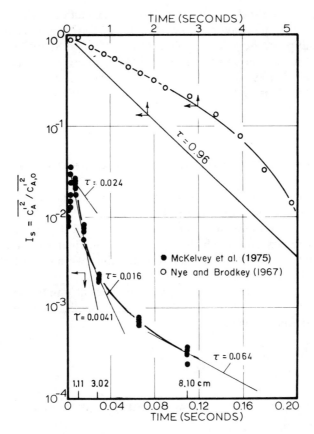

Figure 8 Comparison with theory for mixing (reactor geometry and pipe data).

fraction conversion (F) by

$$I_s = (1 - F)^2 \qquad (19)$$

The agreement is excellent for all times after the coalescence plane, and since Toor's theory does not apply to coarse nonhomogeneous fields, agreement could not be expected before this position.

The results for mixing and its prediction in a variety of geometries have been recently reviewed by Brodkey[2]. It is interesting to note the range of time constants predicted and compared to experiments. The slow mixing at the center line of a pipe is characterized by a mixing time constant of about 1 s. For stirred tank mixing, the constant was about 0.02 s close to the impeller and 0.2 s away from

the impeller. Near the inlet of the reactor feed module, the constant
was 0.0044 s, and away from the head it was 0.067 s. For gas reaction
in a small pipe, the values ranged from 2×10^{-4} to 8×10^{-4} s. Nearly
a 10,000 fold range in mixing times is adequately estimated, if one
has some idea as to the proper value of the injection scale to be
used. The measured values and the corresponding estimates are
given in Ref. 2.

6. EFFECTS OF MIXING ON CHEMICAL REACTION

In earlier sections, the turbulence in a specific reactor, the mixing
and its dependency upon the turbulent field, and the relation of the
mixing to very rapid chemical reactions were all discussed in some
detail. The problem in which both mixing and chemical kinetics are
important contributors to determining the overall reaction rate
remains to be discussed.

The rate term, $\bar{R}_A$, of Eq. (9) contains the information necessary
for the interaction of turbulent mixing and kinetics. For an irreversible reaction, $\bar{R}_A$ is given by

$$-\bar{R}_A = -\frac{d\bar{C}_A}{dt} = k_n \overline{C_A^n} \tag{20}$$

In a first-order reaction, n is unity, and Eq. (20) becomes

$$-\bar{R}_A = -\frac{d\bar{C}_A}{dt} = k_1 \bar{C}_A \tag{21}$$

This first-order expression is a simple linear term and has no
interactive effect on the rate of mixing. The reaction contributes to
the decay of the fluctuation intensity, but does so at exactly the
same rate as it causes a decay in the mean concentration; that is, for
first-order reaction without molecular diffusion (from the equation
for $C_A'^2$)

$$I_s = e^{-2k_1 t} \tag{22}$$

which can be compared to

$$(\bar{C}_A / \bar{C}_{A0})^2 = e^{-2k_1 t} \tag{23}$$

as obtained directly from Eq. (21). From this one can see that mixing has no effect on the conversion of a first-order reaction. The system can best be pictured as a reacting mixture being mixed with a solvent. Diffusivity reduces the concentration fluctuations; superimposed on this is the reduction at a fixed rate by the reaction of both the rms intensity and the mean concentration.

The second-order reaction ($n = 2$) is more complicated. Equation (20) for one component reacting becomes

$$-\bar{R}_A = -\frac{d\bar{C}_A}{dt} = k_2\overline{C_A^{\,2}} = k_2(\bar{C}_A^{\,2} + \bar{C}_A^{\prime 2})$$ (24)

or if it were two components, one obtains

$$-\bar{R}_A = -\frac{d\bar{C}_A}{dt} = k_2\overline{C_A C_B} = k_2(\overline{C_A C_B} + \overline{C_A^\prime C_B^\prime})$$ (25)

Note here that mixing affects $\overline{C_A^{\prime 2}}$ (or $\overline{C_A^\prime C_B^\prime}$) and thus has an interacting nonlinear effect on the reaction. For one component reacting, the fluctuations ($\overline{C_A^{\prime 2}}$) always increase the rates according to Eq. (24). However, when two streams are involved where one contains A and the other B, Toor (21) has shown that the product $\overline{C_A^\prime C_B^\prime}$ is always negative so that the net reaction rate is decreased over what one would expect from an integration of Eq. (20) without consideration of the fluctuating terms.

The prediction of second-order, irreversible chemical reactions where both the kinetics and turbulence are important has been studied in some detail[10, 11, 16, 21–24]. The descriptive equation (9) can be integrated from the inlet if measured velocity data are available (to give $\bar{U}$), if one assumes steady state (i.e., $\partial\bar{C}_A/\partial t = 0$), that the velocity and concentration fields are not correlated (i.e., $\overline{U^\prime C_A^\prime} = 0$), that the diffusion term is small (i.e., $D\nabla^2\bar{C}_A = 0$), and that mixing is not affected by kinetics (Toor's hypothesis) (i.e., so $\overline{C_A^\prime C_B^\prime}$ can be estimated). This latter is adequate for stoichiometric ratios as high as 3.88. For these conditions, Eq. (9) coupled with Eq. (25) for the rate reduces to

$$(\bar{U}\cdot\nabla)\bar{C}_A = -k_2(\bar{C}_A\bar{C}_B + \overline{C_A^\prime C_B^\prime})$$ (26)

or for one direction (x) only

$$\bar{U}_x\frac{\partial\bar{C}_A}{\partial x} = -k_2(\bar{C}_A\bar{C}_B + \overline{C_A^\prime C_B^\prime})$$ (27)

Since $\bar{C}_B$ can be related to $\bar{C}_A$ by the stoichiometry of the reaction (stoichiometric ratio is β), $\bar{U}_x$ is provided from input data, and $\overline{C'_A C'_B}$ is obtained from the mixing (I_s of Eq. 3), Eq. (27) can be easily integrated for $\bar{C}_A$ as a function of position (distance from the mixing head) by, for example, a Runge–Kutta method. The diffusion term could have been retained, since it makes the integration only slightly more complicated, but for liquids it is very small and not needed. Patterson[22] has shown that an interaction between segregation and reaction rate exists which could be important in some instances.

The velocity measurements actually used were made in air and scaled to the temperature conditions actually used to measure the kinetics. It may be well to emphasize once again that the calculation used in this work used actual measurements for the velocity and measured values of I_s. Everything is defined, and the course of the conversion ($\bar{C}_A/\bar{C}_{A0}$) versus distance from the reactor head is computed directly. There are no phenomenological models or assumptions such as a preliminary ideal mixer step as Mao and Toor[11] required. Thus this work is a more stringent test of the hypothesis made by Toor and is an extension to larger values of the stoichiometric ratio (β) than Mao and Toor suggested.

Figure 9 shows three of five experiments. The stoichiometric ratio varied from 1.26 to 3.88. The solid lines are the calculated results from Eq. (27). The experiments reported by Mao and Toor[11] are

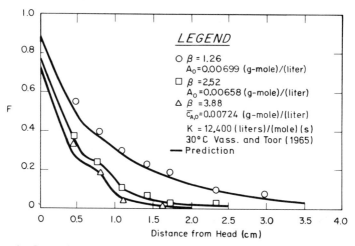

Figure 9   Comparison between observed and estimated conversion for intermediate rate reaction studied by Vassilatos and Toor.

expected to be considerably more reliable than those of Vassilatos and Toor[10]. Our calculation procedure was identical to that just described, and several of the results are shown in Fig. 10. We should note that if one attempts our calculation with the bulk average velocity rather than the actual velocity, the results are unsatisfactory. The overall prediction is not perfect for high conversion ($\beta = 3$) in Fig. 10. The deviation between the prediction and experiment is large when the reaction is nearly complete. This, however, must be blamed on the data, since the last two points are negative, that is, greater than 100% conversion.

In addition to the intermediate rate reaction measurements, Vassilatos and Toor[10] measured relatively slow reactions. They were able to predict the experimental results for the slowest rate by assuming homogeneity. They were unsuccessful in predicting the somewhat faster reaction rate case by a similar assumption. Figure 11 presents some of the computed results for this case. The value of n, which varies from 1 to 2, depends on the relative concentration of ammonia and carbon dioxide. In spite of the uncertainty in n, the theory adequately predicts the slow reaction results.

From this work one may conclude that in a well-defined turbulent field, estimates can be made of the expected mixing, and this in turn can be used to predict chemical reaction with simple kinetics. The results are significant for turbulent chemical reactor design.

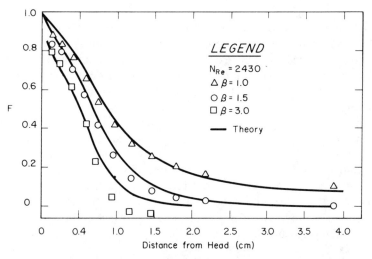

Figure 10   Comparison between observed and estimated conversion for intermediate rate reaction studied by Mao and Toor.

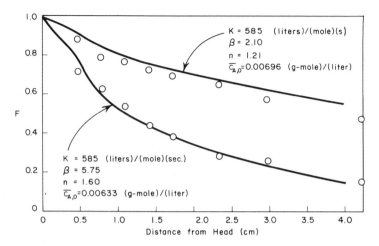

Figure 11   Comparison between observed and estimated conversion for slow rate reaction studied by Vassilatos and Toor.

The work can be extended to more complex reactions[4] such as the series-parallel reaction

$$A + B \xrightarrow{\ k_1\ } P$$

$$B + P \xrightarrow{\ k_2\ } S$$ (28)

The variety of effects that can be encountered in such cases (Bourne and Toor[25]) has resisted a comprehensive treatment.

For the series-parallel reaction (and for other complex reactions), Toor's hypothesis can be extended to apply to the second reaction and a second hypothesis (mixing closure) can be made to relate the mixing for the second reaction to that for the first. Toor's hypothesis, applied to both reactions, can help to simplify the expressions, but still the mixing closure is required. It is, of course, an extension of Toor's hypothesis to apply it independently to the second reaction and thus further assume that the consumption of B in the second reaction has no effect on $\overline{C_A' C_B'}$ that appears in the first equation. For the series-parallel reaction, the mixing for the first reaction involves $\overline{C_A' C_B'}$ and that for the second involves $\overline{C_P' C_B'}$. The mixing closure hypothesis, which allows solution of the equation, is

$$\overline{C_P' C_B'} = \overline{C_A' C_B'} \, (\bar{C}_P / \bar{C}_A)$$ (29)

Results of using this hypothesis will be illustrated; in particular, the effect on selectivity will be discussed.

A simplistic reaction cell model confirmed that Eq. (29) might be valid. Further calculations based on simulating the reactions between cells with random concentrations show that the closure is valid to at least within a few percent when the kinetic coefficients k_1 and k_2 are equal. Thus, Eq. (29) should be explored to see the possibilities of its application.

The reactions are given in Eq. (28) with initial conditions $\bar{C}_{A0}/\bar{C}_{B0} = \beta$, $\bar{C}_{P0} = 0$, $\bar{C}_{S0} = 0$ where β is the stoichiometric ratio. The definitions for the mass fractions are $F_A = \bar{C}_A/\bar{C}_{A0}$, $F_B = \bar{C}_B/\bar{C}_{B0} = \beta\bar{C}_B/\bar{C}_{A0}$, $F_P = \bar{C}_P/\bar{C}_{A0}$, and the resulting equations are (neglecting the diffusion terms as in Eq. (27))

$$\dot{F}_A = -\frac{k_1\bar{C}_{A0}}{\beta}(F_A F_B - I_S)$$

$$\dot{F}_B = -k_1\bar{C}_{A0}(F_A F_B - I_S) - k_2\bar{C}_{A0}(F_B F_P - I_S F_P/F_A)$$

$$\dot{F}_P = +\frac{k_1\bar{C}_{A0}}{\beta}(F_A F_B - I_S) - \frac{k_2\bar{C}_{A0}}{\beta}(F_B F_P - I_S F_P/F_A)$$

$$\dot{F}_S = +\frac{k_2\bar{C}_{A0}}{\beta}(F_B F_P - I_S F_P/F_A)$$

(30)

when I_S is defined by Eq. (3) and we substituted the closure for $\overline{C'_B C'_P}$, as given in Eq. (29). In preliminary calculation, the parameters $\bar{C}_{A0}$, β and k_1 were chosen to reproduce the calculations in reference (1) when $k_2 = 0$.

I_s is assumed unaffected by either chemical reaction, that is, that I_s remains unchanged when k_2 varies. This sets I_s as that for pure mixing as measured either by mixing or by fast reactions. The integration of the four coupled eqs. (30) presents no difficulty. Several results are illustrated in Fig. 12. Until actual measurements are available to compare with these predictions, we can only remark that nothing unexpected or puzzling happens. The $\bar{C}_P/\bar{C}_S$ curves are monotone as expected for all combinations of the parameters that were considered.

The generally satisfactory nature of these results answers some questions but clearly stimulates others. The mixing closure and the independence of I_s on the reactions (extended Toor's hypothesis) appear to be good working tools.

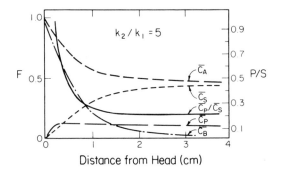

Distance from Head (cm)

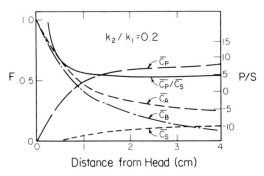

Distance from Head (cm)

Figure 12   Evolution of the concentrations in the tubular reactor for two combinations of the kinetic coefficients.

7. NOTATION

C_A instantaneous concentration fraction, dimensionless
$\bar{C}_A$ average concentration fraction, dimensionless
C'_A instantaneous fluctuation concentration fraction, dimensionless
D diffusion coefficient or molecular diffusivity
F_A fraction converted, $\bar{C}_A/\bar{C}_{A0}$
$g_s(r)$ Eulerian concentration correlation, dimensionless
I_s intensity of segregation, dimensionless
k_n reaction rate constant
L_s scale of segregation
M mass

N_{Sc} Schmidt number (ν/D)
n order of reaction
p pressure
P power
$\mathbf{r}$ distance vector in space
R_A reaction rate term for species A
t time
$\mathbf{x}$ position in space
$\mathbf{U}$ vector velocity
∇ del, vector operator

Greek Symbols

β stoichiometric ratio
ϵ turbulent kinetic energy dissipation
ν kinematic viscosity or momentum diffusivity
η efficiency
σ dimensionless length ($\approx \sqrt{\overline{U'^2}t/L_S}$)
τ mixing time constant

Superscripts

fluctuating value, model system

Subscripts

0   initial value
overbar   time average value at a point

REFERENCES

1. McKelvey, K. N., Yieh, H.-C., Zakanycz, S., and Brodkey, R. S. (1975). Turbulent motion, mixing, and kinetics in a chemical reactor configuration, *AIChE J.*, **21**, 1165.
2. Brodkey, R. A. (1975). Turbulent motion and mixing, in *Turbulence in Mixing Operations* (Ed. R. S. Brodkey,), chap. 2, Academic Press, New York.
3. Brodkey, R. S. (1978). Turbulent motion, mixing, and kinetics, *Proc. Levich Birthday Conf.*, Oxford, England, p. 289.
4. Brodkey, R. S., and Lewalle, J. (1985). Reactor Selectivity Based on First-Order Closures of the Turbulent Concentration Equations, *AIChE J.*, **31**, 111.
5. Taylor, G. I. (1921). *Proc. London Math. Soc.*, **20**, 196.
6. Danckwerts, P. V. (1958). *Chem. Eng. Sci.*, **8**, 93.
7. Danckwerts, P. V. (1953). *Appl. Sci. Res.*, **A3**, 279.
8. Lee, J., and Brodkey, R. S. (1964). Turbulent motion and mixing in a pipe, *AIChE J.*, **10**, 187.

9. Rao, M. A., and Brodkey, R. S. (1972). Continuous flow stirred tank turbulence parameters in the impeller stream, *Chem. Eng. Sci.*, **27**, 137: (1972). *ibid*, Mixing and recycle for analysis of a continuous flow stirred tank, *ibid*, 2199.

10. Vassilatos, G., and Toor, H. L. (1965). *AIChE J.*, **11**, 666.

11. Mao, K. W., and Toor, H. L. (1971). Second-order chemical reactions with turbulent mixing, *Ind. Eng. Chem. Fund.*, **10**, 192.

12. Nye, J. O., and Brodkey, R. S. (1967). Light probe for the measurement of turbulent concentration fluctuations, *Rev. Sci. Insts.*, **38**, 26.

13. Batchelor, G. K. (1959). Small-scale variation of convected quantities like temperature in turbulent fluid, *J. Fluid Mech.*, **5**, 113.

14. Beek, J., Jr., and Miller, R. S. (1959). *Chem. Eng. Prog. Symp. Ser. No. 25*, **55**, 23.

15. Corrsin, S. (1951). On the spectrum of isotropic temperature fluctuations in an isotropic turbulence, *J. Appl. Phys.*, **22**, 469.

16. Toor, H. L., and Singh, M. (1973). The effect of scale on turbulent mixing and on chemical reaction rates during turbulent mixing in a tubular reactor, *Ind. Eng. Chem. Fund.*, **12**, 448.

17. Corrsin, S. (1957). Simple theory of an idealized turbulent mixer, *AIChE J.*, **3**, 329.

18. Corrsin, S. (1964). The isotropic turbulent mixer: part II, arbitrary Schmidt number, *AIChE J.*, **10**, 870.

19. Brodkey, R. A., McKelvey, K. N. Hershey, H. C., and Nychas, S. G. (1978). Mass transfer at the wall as a result of coherent structures in a turbulently flowing liquid, *Int. J. Heat Mass Trans.*, **21**, 593. See also the recent papers by Hanratty (e.g., Campbell, J.A., and Hanratty (1983). Mechanism of turbulent mass transfer at a solid boundary, *AIChE J.*, **29**, 221.

20. Nye, J. O., and Brodkey, R. S. (1967). The scalar spectra in the viscous-convective subrange, *J. Fluid Mech.*, **29**, 151.

21. Toor, H. L. (1962). Mass transfer in dilute turbulent and non-turbulent systems with rapid irreversible reactions and equal diffusivities, *AIChE J.*, **8**, 70.

22. Patterson, G. K. (1973). Model with no arbitrary parameters for mixing effects on second-order reaction with unmixed feed reactants, *Proc. ASME Mixing Symp.* Altanta, GA.

23. Patterson, G. K., Lee, W. C. and Calvin, S. J. (1977). Measurement and numerical modelling of turbulent scalar mixing and reaction in coaxial jets, *Symp. on Turbulent Shear Flow*, Penn. State University.

24. Patterson, G. K. (1979). Closure approximations for complex, multiple reactions, *2nd. Symp. on Turbulent Shear Flow*, Imperial College, London.

25. Bourne, J. R., and Toor, H. L. (1977). Simple criteria for mixing effects in complex reactions, *AIChE J.*, **23**, 602.

Chapter 3

Modelling of Turbulent Reactors

GARY K. PATTERSON

Department of Chemical Engineering, University of Arizona, Tucson, AZ 85721, USA

1. INTRODUCTION

Once a basic understanding of turbulent mixing has been obtained, application of that knowledge to effects on chemical reactions may be attempted. How turbulent mixing affects chemical reactions is dependent on the types of reactions and how the reactants are fed to the mixing zone. It is generally recognized that first-order, irreversible reactions are *not* affected by local turbulent mixing, but *are* affected by convection patterns. The convection patterns in the mixing vessel determine the residence-time distribution of the fluid, from which the first-order reaction conversion may be calculated without concern for the local turbulent mixing. For a complete treatment of the first-order reaction problem with arbitrary convection patterns, see Levenspiel[1]. There the use of the residence-time distribution (RTD) is developed. Nauman[2] has reviewed the use of RTD methods in detail.

Second-order chemical reactions are affected by the rate of local turbulent mixing in ways which depend on whether the reactants are premixed. If they are premixed, the extent of reaction conversion is reduced by the combination of circulatory convection and local turbulent mixing, commonly called backmixing and micromixing, respectively. For the extremes of micromixing, complete and none, the extent of reaction conversion may be computed knowing only the RTD. If the RTD corresponds to that for an ideal stirred tank, the case for no local turbulent mixing is frequently called

59

macromixing. Both analytical and graphical methods for handling these cases are presented by Levenspiel[1].

If the degree of local turbulent mixing is intermediate, both the cases of premixed reactant feed and non-premixed reactant feeds for second-order reactions become much more complicated. In order to account for all the effects of reactant mixing and mixing of reactants with products, then convection, turbulent diffusion, and local turbulent mixing must be modeled. Besides the simultaneous large scale turbulent diffusion and small scale local turbulent mixing processes, indicated by changes in the average concentration gradient and the fluctuating concentration (segregation), there are also the creation of segregation by large scale turbulent diffusion, the reduction of segregation—as well as average concentration of reactants—by chemical reaction, the generation of both large scale and small scale temperature gradients, and effects of both temperature and composition on density and viscosity. The interaction of the reaction with the local turbulent mixing must be accounted for in the presence of all these other effects.

So far, only first-order and second-order irreversible chemical reactions have been discussed. The introduction of more complicated reactions and combinations of reactions makes the problem even more difficult. In what follows, methods will be described for modelling steady-state and unsteady-state reactors with and without significant turbulent diffusion of large scale, for simple second-order irreversible reactions and for complex combinations of reactions (even polymerizations), and for some cases of two-phase transport as part of the reaction system. The methods to be described are categorized as cell-balance models, coalescence-dispersion models, and finite-difference solution of balance equations, both hydrodynamic and mixing.

2.   CHOICE OF MIXING MODEL COMPLEXITY

The least complex model one might conceive for a mixing vessel with chemical reaction would be a lumped system in which the effect of the degree of mixing on chemical conversion is correlated with externally observable variables. Those variables could be the Damköhler number based on chemical rate constant and average residence time, some measure of mixing intensity such as impeller power per unit volume or mass, and geometry factors such as impeller to tank diameter ratio or tank diameter to height ratio or static mixer element size to pipe size ratio. Such an approach may give successful narrow range correlations for single parameter variations for a specific chemical and mixer system. Any chance of a very general result, however, would seem slim.

The incorporation of residence time distribution information leads to the possibility of modelling the effects of the *convection patterns* in the vessel on the chemical reaction. As mentioned above this technique is limited to first-order reactions for a general result and to the extremes of micromixed and segregated for other reaction orders. Also the treatment of non-premixed feeds and multiple reactions requires strong assumptions about the mixing environment.

The treatment of the mixed reactor using RTD information is essentially a Lagrangian approach—the history of each fluid element is analyzed. If the degree to which each fluid element mixed with other fluid elements throughout its history in the mixing vessel were determined, it would then be possible to model the mixing effects on any chemical reaction. This is generally not possible, but an approximation may be made using the coalescence–dispersion model.

In the coalescence–dispersion approach, packets of fluid entering the vessel are moved through the vessel in such a way as to approximate the convection pattern in the real case. In contrast to the segregated model based on RTD concepts the packets are allowed to mix with one another according to a set of rules designed to simulate the mixing rates that would actually occur. Since each packet maintains its identity throughout its passage through the reaction vessel, the model is Lagrangian, but Eulerian information on the distribution of concentrations may be obtained at each time interval. Coalescence–dispersion is easily used to simulate unsteady-state processes and works well for complex chemistry. Nonpremixed feed streams are no problem.

Coalescence–dispersion modelling may be done by actually simulating the events for each fluid packet or by using mathematical functions describing probabilities of coalescence directly. In either case such modelling consumes much computer time. Much more efficient computation is achieved if an Eulerian model is conceived in which balance equations about macroscopic elements or cells of the reaction vessel are written. If the segregation, as well as molar concentration, is regarded as a property of the fluid, then balance equations are needed for all cells in the vessel for each of these two properties plus any others of importance which may vary with reaction conversion, such as temperature, viscosity, or density. If the equations are solved in the unsteady-state mode, a good code for ordinary differential equations is necessary to solve the set. Stiff sets may result which require special methods for handling them.

If the balance equations are solved for a steady-state solution and they are linear, then simple matrix inversion suffices. If, however, any equation set is nonlinear, a Newton method for solution is necessary. Even so, the approach is much faster than the coalescence–dispersion simulation method.

The cell-balance model may be used where good knowledge is available for convection patterns in the vessel from which to determine flow rates into and out of cells and where some knowledge of the local mixing rates are known. Where this is not known, or where greater generality is desired, or where substantial turbulent diffusion exists, a finite difference (or finite element in some cases) model might be best. The cell-balance model is, of course, an approximation to the finite difference model and gives the same results when the cell size is small enough to give the same resolution. An advantage of the finite difference technique, however, is that momentum balance equations and mass balance equations with significant turbulent diffusion may be solved. The computer time required depends on the number of balance equations and the number of grid points in the simulation. Methods for unsteady-state problems have been formulated and used in finite difference modelling, but steady-state problems are most often solved using this approach.

There are various levels of complexity that may be used for finite difference modelling to obtain the distributions of fluid velocities, turbulence intensities, rates of mixing (or turbulence energy dissipation rates and turbulence scales), segregations, thermal energies, and concentrations for each component in the mixing vessel. The hierarchy of such models for the hydrodynamic part (velocities, turbulence, energy dissipation, and scales) has been adequately discussed by Launder and Spalding[13]. In order to model mixing of chemical components, it is necessary to use a model of sufficient complexity that the local mixing rate information can be derived. Usually this means a hydrodynamic model which includes balance equations for mass, momentum, turbulence energy, and turbulence energy dissipation rate, even though simpler finite difference models exist. The reason for this will be shown in a later section.

The model hierarchy for the mixing part will be discussed in detail in a later section. The complexity necessary depends on whether the reaction rate is fast enough to assume equilibrium of reactants and products, whether significant temperature fluctuations are generated, and whether more than one reaction is taking place.

2.1. Interaction of Local Turbulent Mixing and Large-Scale Turbulent Diffusion

If two miscible fluids are mixing in a turbulent flow with an initial segregation of a relatively small scale, *two processes* combine to cause the two fluids to mix. One is the stretching of fluid elements by the turbulence causing the concentration gradients of the two fluids to increase and also drawing the regions of the two fluids closer together. The other process is the molecular diffusion which ultimately mixes the two fluids on a molecular level and which is greatly enhanced by the turbulent stretching. The process may be modeled one-dimensionally either timewise[3] or spatially[4].

If the two fluids enter the mixing region in such a way that they fill two separate but adjacent regions so that fluid dispersion over relatively large distances must occur to bring about mixing, then a third process—turbulent diffusion—must take place. Turbulent diffusion may be treated as a large scale phenomenon. The process must now be modeled as a two-dimensional[5,6] or a three-dimensional system.

The mixing and diffusion phenomena described above are purely physical processes. If the fluids involved react chemically under the conditions of mixing, considerable interaction between the two processes may occur. This is particularly true if the reaction or reactions occur at about the same rate as the mixing, so that concentration gradients are increased by reaction *and* reaction rate is increased by mixing rate. Very important interactions occur if more than one reaction occurs simultaneously, the selectivity to specific products being a strong function of mixing rate.

3. BALANCE EQUATIONS FOR HYDRODYNAMICS AND MIXING

The most fundamental approach to modelling the interactions of mixing, turbulent diffusion, convection, and chemical reaction is through the formulation and solution of the differential balance equations which describe the process. In this paper the general formulation will be followed by various simplifications and variations of modelling method.

The modelling of mixing without large scale turbulent diffusion both with and without chemical reaction has been extensively studied. Both hydrodynamic[3,4] and coalescence–dispersion (c–d)[7,8] methods have been used to model mixing successfully in such one-dimensional systems.

Two-dimensional mixing systems are considerably more difficult to model, because of the turbulent diffusion which affects all the dependent variables—which are concentration, velocity, turbulence energy, turbulence energy dissipation rate, and segregation if a full finite difference model is used. Such two-dimensional systems have been successfully modelled without reaction for the round free jet[9], the confined round jet[10,11], and the coaxial free jet[6] as well as some other more complex geometries. The round free jet[11] and the coaxial free jet[6] have also been modelled for second-order reaction occurring between the two mixing fluids. In Ref. 11 chemical equilibrium was assumed; in Ref. 6 it was not. All the modelling efforts above were finite difference models based on the use of the $k - \epsilon$ model[12] for the hydrodynamics (to be described later).

The differential balance equations that *must* be included in a general formulation are those for overall mass, momentum, component mass (concentration), and thermal energy. In order to facilitate the solution of the momentum balance, balances for the turbulence energy and turbulence energy dissipation rate or turbulence length scale are frequently added. That forms the minimum set for the case of *two-equation* (beyond the momentum balance) simulation (turbulence energy and turbulence energy dissipation rate) of the hydrodynamics (see Ref. 13). More balance equations may be necessary if a full Reynolds stress model for the hydrodynamics is used: three Reynolds stress balance equations replace the turbulence energy equation.

In order to improve the mixing model, it is necessary to add balance equations for the segregation of each component. An alternate scheme with equations for the velocity-concentration correlations has also been employed. Hill[14] has given a useful review of the various modelling methods commonly used. If temperature fluctuations are important, it may also be necessary to include balance equations for temperature-velocity and temperature-concentration interactions (correlations).

3.1. Hydrodynamic Modelling of Turbulence

Here the two-equation approach (two balance equations beyond the continuity and momentum equations) using equations for turbulence energy and dissipation will be described. The overall mass balance equation, frequently called the continuity equation, is as follows:

$$\frac{D\rho}{Dt} = -\rho \left(\frac{\partial U_i}{\partial x_i}\right) \tag{1}$$

where $D/Dt = \partial/\partial t + U_j \partial/\partial x_j$.

The momentum balance equation for instantaneous variables (constant density and viscosity) is:

$$\frac{DU_i}{Dt} = -\frac{1}{\rho}\frac{\partial \rho}{\partial x_i} + \nu \left(\frac{\partial^2 U_j}{\partial x_i \partial x_j}\right) + g_i \tag{2}$$

After Reynolds averaging and rearrangement:

$$\frac{D\bar{U}_i}{Dt} = -\frac{1}{\rho}\frac{\partial \bar{p}}{\partial x_i} + \nu \left(\frac{\partial^2 \bar{U}_i}{\partial x_j \partial x_j}\right) - \overline{\rho u_i u_j} + \bar{g}_i \tag{3}$$

In order to model the momentum balance equation based on velocity gradients and turbulent kinematic viscosities, Eq. (3) is rewritten in two dimensions as follows:

$$\frac{D\bar{U}_1}{Dt} = -\frac{1}{\rho}\frac{\partial \bar{p}}{\partial x_1} + \frac{\partial}{\partial x_1}\left(\frac{\nu_e + \nu}{1}\frac{\partial \bar{U}_1}{\partial x_1}\right) + \frac{\partial}{\partial x_2}\left(\frac{\nu_e + \nu}{1}\frac{\partial \bar{U}_1}{\partial x_2}\right) \tag{4a}$$

$$\frac{D\bar{U}_2}{Dt} = -\frac{1}{\rho}\frac{\partial \bar{p}}{\partial x_2} + \frac{\partial}{\partial x_1}\left(\frac{\nu_e + \nu}{1}\frac{\partial \bar{U}_2}{\partial x_1}\right) + \frac{\partial}{\partial x_2}\left(\frac{\nu_e + \nu}{1}\frac{\partial \bar{U}_2}{\partial x_2}\right) \tag{4b}$$

The turbulent kinematic viscosity ν_e is a momentum diffusivity and the body force term g_i has been dropped.

If Eq. (2) is multiplied by u_i, the fluctuating velocity in the i direction, each of the resulting three equations is Reynolds averaged, then the corresponding terms are summed, the turbulence energy equation results:

$$\frac{Dk}{Dt} = -\frac{\partial}{\partial x_i}\left(\overline{u_i k} + \frac{1}{\rho}\overline{u_i p}\right) - \overline{u_i u_j}\frac{\partial \bar{U}_i}{\partial x_j} - \nu \overline{\left(\frac{\partial u_i}{\partial x_j}\right)^2} \tag{5}$$

$$\text{(diffusion)} \quad \text{(production)} \quad \text{(dissipation)}$$

k represents the turbulent kinetic energy, $\frac{1}{2}(\overline{u_i^2})$. Again, in order to model the turbulence energy equation based on turbulence energy gradients, the equation is rewritten as follows:

$$\frac{Dk}{Dt} = \frac{\partial}{\partial_1}\left(\frac{\nu_e + \nu}{\sigma_k}\frac{\partial k}{\partial x_1}\right) + \frac{\partial}{\partial x_2}\left(\frac{\nu_e + \nu}{\sigma_k}\frac{\partial k}{\partial x_2}\right)$$

(diffusion)

$$+ \nu_e\left\{\left(\frac{\partial \bar{U}_1}{\partial x_2} + \frac{\partial \bar{U}_2}{\partial x_1}\right)^2 + 2\left(\frac{\partial \bar{U}_i}{\partial x_i}\right)^2\right\} - \epsilon \qquad (6)$$

(production)   (dissipation)

The diffusivity for the turbulence energy is given by the ratio of the momentum diffusivity and a constant σ_k of the order one. The term for turbulence energy production rate is frequently reduced to ν_e $[(\partial \bar{U}_1/\partial x_2) + (\partial \bar{U}_2/\partial x_1)]^2$, and the term for dissipation rate is reduced to one dependent variable ϵ which must be modelled by a balance equation.

The balance equation for the turbulence energy dissipation rate may be obtained by taking the derivative of Eq. (2) with respect to x_l, multiplying through by $2\nu(\partial u_i/\partial x_l)$, then Reynolds averaging. After some rearrangement the dissipation term as in Eq. (5) appears as the dependent variable. For simplicity the complete equation is not shown. The model balance equation is given in the same form as the model equation for turbulence energy:

$$\frac{D\epsilon}{Dt} = \frac{\partial}{\partial x_1}\left(\frac{\nu_e + \nu}{\sigma_\epsilon}\frac{\partial \epsilon}{\partial x_1}\right) + \frac{\partial}{\partial x_2}\left(\frac{\nu_e + \nu}{\sigma_\epsilon}\frac{\partial \epsilon}{\partial x_2}\right)$$

$$+ C_1\nu_e\left\{\left(\frac{\partial \bar{U}_1}{\partial x_2} + \frac{\partial \bar{U}_2}{\partial x_1}\right)^2 + 2\left(\frac{\partial \bar{U}_i}{\partial x_i}\right)^2\right\}\epsilon/k - C_2\epsilon^2/k \qquad (7)$$

(production)   (dissipation)

The production and dissipation terms "model" the actual collection of terms in the complete equation. Such an approximation is necessary in order to force closure of the set of differential balance equations so that a solution can be found. Again, the term $2(\partial \bar{U}_i/\partial x_i)^2$ is frequently dropped.

In Eqs. (1)–(7) the values of the "constants" are somewhat dependent on the flow geometry, but for most cases the following are recommended by Jones and Launder[12]: $C_1 = 1.44$, $C_2 = 1.92$, $\sigma_k = 1.0$, $\sigma_\epsilon = 1.3$. The turbulent kinematic viscosity $\nu_e = C_D k^2/\epsilon$, where $C_D = 0.09$.

These equations constitute the k-ϵ two-equation model of turbulence and are only one of several methods of modelling the hydrodynamics of turbulent flows. The well known Prandtl mixing-length and von Karman similarity models for the turbulent (eddy) viscosity lead to zero-equation models; that is, zero differential equations beyond the continuity equation and the momentum balance. There are several one-equation models which are useful for boundary layer flows[13]. They are usually not useful for the mixing problems considered in this paper, because most mixing is done in complex geometries, for which two-equation models are more effective. Other two-equation models are the $k - l$ model, the $k/l^2 - k$ model, the $k^{3/2}/l - k$ model and the $k^{1/2}/l - k$ model. Launder and Spalding[13] state that the three most thoroughly tested ($k - \epsilon$, $k - l$, and $k/l^2 - k$) are essentially equivalent. The relationship for the length scale $l = k^{3/2}/\epsilon$ is assumed to hold in each of the models.

More complex modelling methods, such as the Reynolds stress models, are available for hydrodynamic modelling of turbulent flows. Unfortunately, the number of equations to solve is larger, which is a great disadvantage, since the mixing model also requires several equations. Also, as is shown below, the $k - \epsilon$ model for hydrodynamics fits perfectly with the model for mixing.

3.2. Modelling Turbulent Mixing with Chemical Reaction

The differential balance equation for a component in a turbulent mixing flow may be written as:

$$\frac{DC_i}{Dt} = D_i \left(\frac{\partial^2 C_i}{\partial x_n^2} \right) + R_i \tag{8}$$

D_i is the diffusivity of component i in a mixture. If the source term R_i is modelled by an irreversible second-order reaction between components A_i and A_j, then:

$$R_i = -KC_iC_j \qquad \text{for } i \neq j \tag{9}$$

K is the rate constant for $A_i + A_j \rightarrow A_k$.

If C_i, U_n and K are divided into average and fluctuating quantities, such as $\bar{C}_i$ and c_i, then the resulting equation is Reynolds averaged to yield:

$$\frac{D\bar{C}_i}{Dt} = D_i \left(\frac{\partial^2 \bar{C}_i}{\partial x^2} \right) - \frac{\partial}{\partial x_n} \left(\overline{u_n c_i} \right) - \bar{K}(\bar{C}_i\bar{C}_j + \overline{c_i c_j}) - \overline{k c_i}\, \bar{C}_j - \overline{k c_j}\, \bar{C}_i - \overline{k c_i c_j} \tag{10}$$

If the temperature is constant or if temperature fluctuations are not highly correlated with concentration fluctuations, the last three terms of Eq. (10) involving k may be dropped. Throughout the remainder of this article that will be done, because the modelling of such temperature fluctuation effects is not yet well developed.

The simplified form of Eq. (10) may be rewritten in the two-dimensional form of Eqs. (4), (6), and (7) as follows:

$$\frac{D\bar{C}_i}{D_t} = \frac{\partial}{\partial x_1}\left(\frac{\nu_e + \nu}{\sigma_{\bar{c}}}\frac{\partial \bar{C}_i}{\partial x_1}\right) + \frac{\partial}{\partial x_2}\left(\frac{\nu_e + \nu}{\sigma_{\bar{c}}}\frac{\partial \bar{C}_i}{\partial x_2}\right) - \bar{K}(\bar{C}_i\bar{C}_j + \overline{c_ic_j}) \quad (11)$$

This is the model component mass balance equation. The last term is later given as $P_{\bar{c}_i}$.

If Eq. (8) is multiplied by c_i then Reynolds averaged, the following transport (balance) equation for segregation $\overline{c_i^2}$ occurs:

$$\frac{D\overline{c_i^2}}{Dt} = -\frac{\partial}{\partial x_n}\overline{(u_nc_i^2)} - 2\overline{c_iu_n}\frac{\partial \bar{C}_i}{\partial x_n} - 2D_i\overline{\left(\frac{\partial c_i}{\partial x_n}\right)^2} - \bar{K}(\bar{C}_i\overline{c_ic_j} + \bar{C}_j\overline{c_i^2} + \overline{c_i^2c_j})$$

(turbulent   (production)·(dissipation)   (reaction effect)
diffusion)

(12)

Just as in the hydrodynamic part a number of terms must be modelled in order to produce closure of the set of differential balance equations. Other equations may be generated, such as a balance equation for $\overline{u_nc_i}$, but closure must always be approximated at some level. Here, closure with the segregation equation will be shown.

It is necessary to model the dissipation term and the production term in order to account for mixing and diffusion without reaction. If second-order reaction occurs, the terms involving $\overline{c_ic_j}$ and $\overline{c_i^2c_j}$ must also be modelled all in terms of $\overline{c_i^2}$ and/or $\bar{C}_i$. Corrsin[3] showed that the rate of mixing of miscible components, hence the rate of dissipation of segregation, may be modelled as a function of a scalar length scale and the rate of turbulence energy dissipation in isotropic turbulence. Since both the final rate of mixing *and* the rate of turbulence energy dissipation are primarily dependent on the smallest scales of turbulence, which tend to be isotropic, Corrsin's model has proven useful even for shear flow turbulence[4, 6, 15]. His relationship may be written:

$$2D_i \left(\frac{\overline{\partial c_i}}{\partial x_n}\right)^2 = 2\overline{c_i^2}/[4(k/\epsilon) + (\nu/\epsilon)^{1/2} \ln N_{Sc}] \tag{13}$$

Here the scalar length scale does not appear because it has been assumed to be approximately equal to $k^{3/2}/\epsilon$, a seemingly crude assumption, but one that seems to work. The usefulness of Eq. (13) is that it involves k and ϵ as hydrodynamic variables, which are provided by solutions to Eqs. (6) and (7). Spalding[5] modelled the segregation dissipation rate as $Cg_2\epsilon\overline{c_i^2}/k$, where Cg_2 was about 0.2. If in Eq. (13) the Schmidt number N_{Sc} is small, the segregation dissipation rate becomes $0.5\epsilon\overline{c_i^2}/k$, which is a higher rate of dissipation than that predicted by Spalding's term.

The turbulent diffusion term is modelled as in all other cases and is shown in the final equation.

The most difficult term to handle is the segregation production term. Spalding[5] modelled the segregation production term analogously to the turbulence energy production term, see Eq. (6). He therefore arrived at the following:

$$2\overline{c_i u_n} \frac{\partial \overline{C_i}}{\partial x_n} = -Cg_1 \nu_e \left(\frac{\partial \overline{C_i}}{\partial x_n}\right)^2 \tag{14}$$

where Cg_1 has a value of approximately 3.

Another approach to the problem is to make use of two ideas: (1) new fluid diffusing into a region by turbulent transport contributes to segregation whether the fluid is perfectly mixed or not and (2) the segregation for unmixed components is given by the product of their average concentrations. A balance equation for imaginary total segregation may be written as follows:

$$-\frac{D(\overline{C_i}(C_T - \overline{C_i}))}{Dt} = 2\overline{c_i u_n} \frac{\partial \overline{C_i}}{\partial x_n} \tag{15}$$

(convection)   (production)

It is assumed that the turbulent diffusion of component i is responsible for the production rate of segregation, and that the rate of imaginary total segregation production is the same. The imaginary total segregation does not diffuse, therefore no diffusion term is involved.

The two-dimensional model equation for segregation with second-order reaction is now:

$$\frac{D\overline{c_i^2}}{Dt} = \frac{\partial}{\partial x_1}\left(\frac{\nu_e + \nu}{\sigma_c^2}\frac{\partial \overline{c_i^2}}{\partial x_1}\right) + \frac{\partial}{\partial x_2}\left(\frac{\nu_e + \nu}{\sigma_c^2}\frac{\partial \overline{c_i^2}}{\partial x_2}\right) + \frac{D(\overline{C_i}(C_T - \overline{C_i}))}{Dt}$$

$$-2\overline{c_i^2}/[4(k/\epsilon) + (\nu/\epsilon)^{1/2}\ln N_{Sc}] - K_{ijk}(\overline{C_i c_i c_j} + \overline{C_j c_i^2} + \overline{c_i^2 c_j}) \qquad (16)$$

The combination of the last three terms is later given as $P_{\overline{c_i^2}}$.

3.3. Closure Approximations for Second-Order Chemical Reactions

Two terms have not yet been modelled: $\overline{c_i c_j}$ and $\overline{c_i^2 c_j}$. The modelling of these terms is really the crux of modelling mixing with second-order chemical reaction. Closure of the component balance and segregation balance equations may be accomplished by a variety of methods: assumption of equilibrium or near equilibrium, higher-order-moment closure, use of probability density functions of the component concentrations, and the use of simple physical models such as coalescence–dispersion, fluid-strand diffusion, eddy mixing, and "exchange-with-the-mean". These latter methods actually are substitutes for the segregation balance equation and not closure methods in the usual sense. Details of each of the closure methods mentioned will be elaborated below.

3.3.1. Equilibrium Assumption
If the chemical reaction is very fast compared to the mixing rate, it may be assumed that any mixed reactants are immediately converted. No rate expression is therefore necessary and the last term of Eq. (11) is related to the rate of mixing. Figure 1 shows typical instantaneous concentration profiles for two mixing components which because of instantaneous reaction remain totally segregated from one another. Brodkey[15] showed that the segregation $\overline{c^2}$ of two totally segregated fluids is the product of their average concentrations $\overline{C_1}\overline{C_2}(= c_1^2 = c_2^2)$. The rate of decrease of $\overline{C_1 C_2}$ due only to reaction may be expressed as follows:

$$-\left(\frac{\partial(\overline{C_1}\overline{C_2})}{\partial t}\right)_r = -\overline{C_1}\left(1 + \frac{\overline{C_2}}{\overline{C_1}}\right)\left(\frac{\partial \overline{C_1}}{\partial t}\right)_r = \left(\frac{\partial \overline{c_1^2}}{\partial t}\right)_m \qquad (17)$$

because $[(\partial\overline{C_1}/\partial t)]_r = [(\partial\overline{C_2}/\partial t)]_r$, when they are due only to reaction. Therefore, the rate of reaction decrease of the component concen-

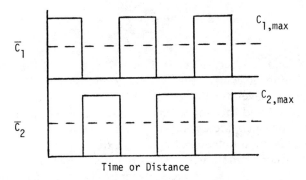

Figure 1   Concentration profiles for completely segrated fluids.

trations becomes:

$$-\left(\frac{\partial \bar{C}_1}{\partial t}\right)_r = -\left(\frac{\partial \bar{C}_2}{\partial t}\right)_r = -\left(\frac{\bar{C}_1}{\bar{C}_1^2 + \bar{C}_1\bar{C}_2}\right)\left(\frac{\partial \overline{c_1^2}}{\partial t}\right)_m \tag{18}$$

With the rate of segregation decrease $-[(\partial \overline{c_1^2}/\partial t)]_m$ given by the Corrsin relation, Eqs. (11) and (16) become:

$$\frac{D\bar{C}_i}{Dt} = \frac{\partial}{\partial x_1}\left(\frac{\nu_e + \nu}{\sigma_{\bar{c}}}\frac{\partial \bar{C}_i}{\partial x_1}\right) + \frac{\partial}{\partial x_2}\left(\frac{\nu_e + \nu}{\sigma_{\bar{c}}}\frac{\partial \bar{C}_i}{\partial x_2}\right)$$
$$- \frac{\bar{C}_i \overline{c_1^2}}{\bar{C}_i^2 + \bar{C}_i\bar{C}_j}\frac{1}{[4(k/\epsilon) + (\nu/\epsilon)^{1/2}\ln N_S]} \tag{19}$$

$$\frac{D\overline{c_i^2}}{Dt} = \frac{\partial}{\partial x_1}\left(\frac{\nu_e + \nu}{\sigma_{\overline{c^2}}}\frac{\partial \overline{c_i^2}}{\partial x_1}\right) + \frac{\partial}{\partial x_2}\left(\frac{\nu_e + \nu}{\sigma_{\overline{c^2}}}\frac{\partial \overline{c_i^2}}{\partial x_2}\right) + \frac{D(\bar{C}_i(C_T - \bar{C}_i))}{Dt}$$
$$- 2\overline{c_i^2}/[4(k/\epsilon) + (\nu/\epsilon)^{1/2}\ln N_{Sc}] \tag{20}$$

The last term of Eq. (16) does not appear in Eq. (20) because with total segregation it becomes zero. Equations (19) and (20) include no effect of reaction on mixing rate, a possible source of error. Since no other moments besides $\overline{c_i^2}$ appear in the equations, they are closed.

In many cases a "conserved scalar" approach to the formulation of balance equations for fast reactions is used. The method has been thoroughly reviewed by Bilger[16]. Gosman et al.[17] have used the mixture fraction, a conserved scalar, to formulate a model for combustion which is useful in many types of problems. The mixture fraction is defined as:

$$\xi = \frac{\overline{C_i^0} - \overline{C_{i2}}}{\overline{C_{i1}} - \overline{C_{i2}}} \tag{21}$$

where 1 and 2 refer to feeds 1 and 2 and $\overline{C_i}$ is the unreacted concentration of component i. For a given value of ξ, which corresponds to a given fraction of feed 1 in the mixture, if the fluids are totally mixed and reaction is complete, the concentrations of reactants and products are as shown on Fig. 2. ξ is a conserved scalar because it is not a function of reaction extent.

For a second-order, irreversible reaction which is very fast as follows $A_1 + nA_2 \rightarrow A_3$, if the mixing is also very fast, the local concentrations of A_1, A_2, and A_3 are functions of ξ and the feed concentrations. If $\overline{C_{12}^f}$ and $\overline{C_{21}^f}$ are zero, (only component 1 in feed 1 and only component 2 in feed 2) the local concentrations with instantaneous reaction are as follows:

$$\overline{C_1} = \xi \overline{C_{11}^f} + n(\xi - 1)\overline{C_{22}^f} \qquad \text{for } \xi > \xi_s; \ \overline{C_1} = 0 \qquad \text{for } \xi \leq \xi_s$$

$$\overline{C_2} = (1 - \xi)\overline{C_{22}^f} - \frac{1}{n}\xi \overline{C_{11}^f} \qquad \text{for } \xi < \xi_s; \ \overline{C_2} = 0 \qquad \text{for } \xi \geq \xi_s$$

Such a formulation does not require the computation of the degree of mixing so along with Eqs. (1), (4), (6), and (7) for the hydrodynamics, the following conserved mass equation may be used to model the chemical reaction in two-dimensions:

$$\frac{D\xi}{Dt} = \frac{\partial}{\partial x_1}\left(\frac{\nu_e + \nu}{\sigma_\xi}\frac{\partial \xi}{\partial x_1}\right) + \frac{\partial}{\partial x_2}\left(\frac{\nu_e + \nu}{\sigma_\xi}\frac{\partial \xi}{\partial x_2}\right) \tag{22}$$

There is no generation term.

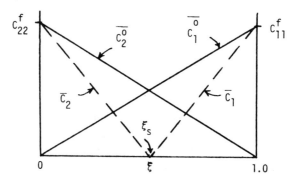

Figure 2   Mixture fraction as a conserved scalar. ξ_s is the mixture fraction corresponding to a stoichiometric mixture.

Spalding[18] has proposed a method for modelling the effect of incomplete mixing when the reaction is much faster than the mixing. In addition to Eqs. (1), (4), (6), (7), and (22) (with a source term $\bar{S}_\xi$), Eq. (20) (with the Spalding term for $\overline{c_i^2}$-production) is used. The level of $\overline{c_i^2}$ is used to determine the values ξ_+ and ξ_-, the upper and lower extents of concentration fluctuations. The source term was given as follows:

$$\bar{S}_\xi = -\{\alpha r_- + (1 - \alpha)r_+\}C_{\mathrm{EBU}}|\partial \bar{U}/\partial x_2| \qquad (23)$$

where

$$\alpha = (\xi_+ - \xi)/(\xi_+ - \xi_-)$$

$$r_+ = (\xi_+ \bar{C}_{1i} - \bar{C}_{i+})(\bar{C}_{i+} - C^*_{i+})/(\xi_+ \bar{C}_{1i} - C^*_{i+})$$

$$r_- = (\xi_- \bar{C}_{1i} - \bar{C}_{i-})(\bar{C}_{i-} - C^*_{i+})(\xi_- \bar{C}_{1i} - C^*_{i-})$$

$\bar{C}_{i+}$ and $\bar{C}_{i-}$ must lie on a straight line with $\bar{C}_i$. The "eddy break-up" constant $C_{\mathrm{EBU}} \approx 0.5$.

The condition under which the equilibrium approach is valid may be expressed in terms of the ratio of the time constants of homogeneous reaction and mixing. The time constant for second-order reaction is $\tau_R = K_{ijk}C_i$ and the time constant for mixing is $\tau_M = 1/[4(k/\epsilon) + (\nu/\epsilon)^{1/2} \ln N_s]$. In order to apply the equilibrium approximation it is necessary that $\tau_M \gg \tau_R$. If $\tau_R \gg \tau_M$, then the mixing rate has no effect on the reaction and it may be treated as a homogeneous reaction. If τ_R and τ_M are of the same order, it is necessary to use a closure method which accounts for the interaction of mixing and reaction.

3.3.2. Higher-Order Moment Closure

If other balance equations for correlations such as $\overline{u_n C_i}$ and for higher-order moments such as $\overline{c_i^3}$ are formed, closure may take place at a higher order than in Eqs. (11) and (16). Typically, at some level the moments are assigned to be zero so that closure is accomplished.

Borghi[19] has investigated the possibilities for higher-order moment closure. He included in the source term for the component mass balance all the terms generated in Eq. (10). By expressing the instantaneous rate constant as an Arrhenius term, $R_i = KC_1C_2T_{\mathrm{ref}} \exp(-T_a/T)$, the source term was eventually put into a series expansion form for which the higher order terms approached zero under certain circumstances. Model equations were necessary for the terms $\overline{h^2}$ (mean square enthalpy fluctuation), $\overline{c_1^2}$, $\overline{c_1 c_2}$, $\overline{c_2 h}$, and $\overline{c_2 \xi}$ for a two-component reacting mixture in addition to the basic hydrodynamic equations, similar to Eqs. (1), (4), (6), and (7), a mass balance equation for component 2, a balance equation for ξ, and a balance equation for $\bar{H}$, the average enthalpy. The complete model was a 13-equation model for only two reacting components, an illustration of the complexity of complete modelling of mixing with reaction.

3.3.3. Probability Density Functions

Since the source terms in the mass balance equations generate all of the difficult, but important, higher order terms which must in some way be modelled, attempts have been made to represent the statistical distribution of the basic variables, which are concentration of each component and temperature (or enthalpy). At the small scales, where most of the important interactions occur in mixing with chemical reaction, the distributions of those quantities become more random than at the large scales where great coherency exists. This has led many investigators to use approximations to the probability density distributions or functions (pdf's) of each of the important quantities to close the equations.

Although very few measurements of pdf's for mixing components have been made, a number of modelers have proposed model pdf's for use in modelling chemical reactions. Lockwood and Naquib[11] have proposed a clipped-normal pdf for the function $p(\xi)$, where $\overline{\phi_1 \phi_2} = \int_0^1 \phi_1(\xi) \phi_2(\xi) p(\xi) \, d\xi$ represents the averaged value of the products of functions of ξ, such as $c_1 c_2$. Any correlation quantity used in the closure of the model equations is thus determined.

The clipped-normal distribution, a spiked distribution, and a more realistic distribution of a concentration are shown in Fig. 3. As can be seen from the figure, the spiked distribution is closer to the perceived realistic distribution than is the clipped-normal distribution. The spiked distribution is derived from a simplified physical model of the concentration profiles expected in a mixing system. Donaldson[20] has called this the "most typical" eddy.

A much simpler version of that approach was formulated previously by Patterson[4] and may be called the "interdiffusion" model. Figures 4 and 5 show the relationships between a crude model of two components diffusing into one-another and the pdf that would result. Application of the interdiffusion model results in the follow-

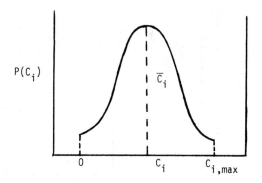

(a)

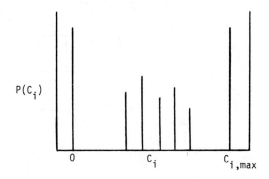

(b)

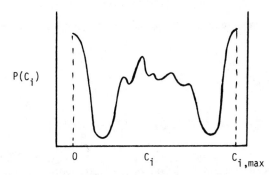

(c)

Figure 3 Probability density functions for mixing-probability of finding a concentration c_i. (a) Clipped-normal distribution function. (b) Spiked distribution function. (c) Typical real distribution function.

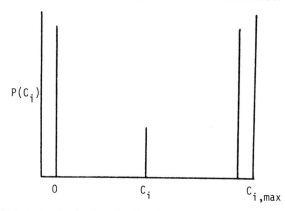

Figure 4   Probability density function for almost segregated fluids with one intermediate concentration.

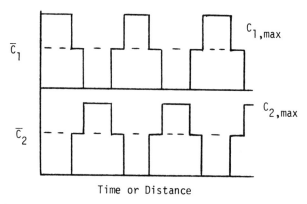

Time or Distance

Figure 5   Concentration profile corresponding to pdf in Fig. 4—Interdiffusion model.

ing relationships:

$$\overline{c_1 c_2} = -\overline{c_1^2}(1 - \gamma)/(\beta(1 + \gamma))$$

$$\overline{c_1^2 c_2} = 2(\overline{c_1^2})^{3/2}\gamma(1 - \gamma)^{1/2}/(\beta(1 + \gamma)^{3/2}) \tag{24}$$

where $\beta = C_{10}/C_{20}$ and $\gamma = (\beta\overline{C_1 C_2} - \overline{c_1^2})/(\beta\overline{C_1 C_2} + \overline{c_1^2})$. Comparisons with experiment showed that best results were obtained when $\overline{c_1^2 c_2}$ was set equal to zero.

Discussion of coalescence-dispersion, fluid-strand diffusion, eddy mixing, exchange-with-the mean, and extension of the interdiffusion model will be presented for complex chemistry below.

4. APPLICATION TO COMPLEX GEOMETRIES

An advantage of the fundamental approach to modelling the hydrodynamics and mixing described so far is its ability to describe even quite complex 2- and 3-dimensional geometries. Examples of such applications of modelling of mixing and reaction are given by Spalding[18] for a bunsen burner, by Ellail et al.[21] for a 3-dimensional combustion chamber, by Lockwood and Nagib[11] in a round, free jet diffusion flame, and by Patterson et al.[6] in a coaxial mixing jet. Not enough is known about the similarity rules which may exist in such complex, reacting systems in order to scale them up or down without resorting to full modelling. The recommended scaleup method in such a case would, therefore, be to confirm the modelling at the pilot scale with experimental measurements, then compute results at larger scales. Since the modelling method above depends on physics as fundamental as possible, some confidence should be possible in the scaleup based on it.

For less complex geometries, which are essentially one-dimensional or where turbulent diffusion is not important, models based on interconnected tanks or cells may be applied. The finite-difference solution of the hydrodynamic equations is not then sought, but flow rates between the tanks is usually determined from experimentally measured flow patterns. Levels and scales of turbulence for each of the tanks are also usually determined from experiments. Only the component mass balance and segregation balance equations are then solved, see Eqs. (11) and (16). Such an approach has been applied to the turbulent tubular reactor[4] and the stirred-tank reactor[25]. Scaleup of the mixing[23] and reaction in such geometries seems to be possible based on the use of the proper dimensionless groups and absolute geometric similarity. Figures 6, 7, and 8 show scaleup correlations, for mixing without reaction and for reaction in tanks stirred by turbines and propellers. Figure 9 shows comparisons of experiment and model for the turbulent tubular reactor.

5. APPLICATION TO COMPLEX CHEMISTRY

5.1. Modelling of Complex Reaction Chemistry with Mixing

Conversion, yield and selectivity in complex (multiple) reactions in a homogeneous mixing environment is determined essentially by the relative rates of reaction and diffusion just as in heterogeneous

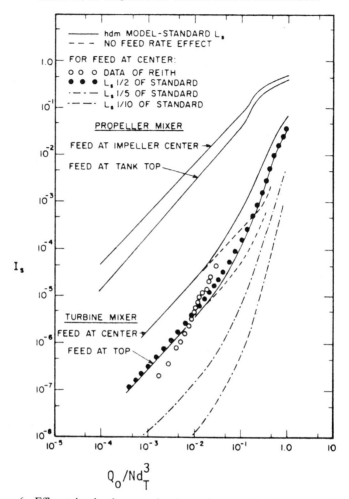

Figure 6 Effluent levels of segregation for tanks stirred by Rushton turbines and propellers. Turbines and propellers centered in tank; impeller diameter d_T one-third of tank diameter D; Q_0/Nd_T^3 is measure of feed rate to pumping rate ratio.

reactions. In a homogeneous mixing environment, however, there does not exist a relatively unchanging solid surface to and from which diffusion occurs and upon which the reaction occurs. In the heterogeneous case, separation of the diffusion and reaction steps is a classical and very successful procedure for most reactions. In the homogeneous mixing case, diffusion and reaction are inseparably interwined, and the concentration gradients in a given fluid locality change constantly, making the application of a quasi-steady state model questionable at best.

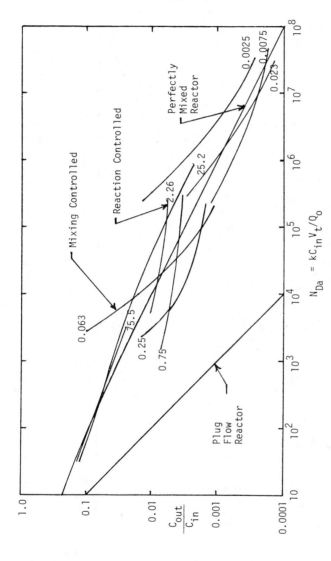

Figure 7 Conversion in the effluent for second-order reactions in a Rushton turbine-stirred tank. Turbine centered in tank; $d_T = D/3$; parameters on curves are mixing intensity given by $(\epsilon/L_s^2)_{center}^{1/3}/kC_{in}$.

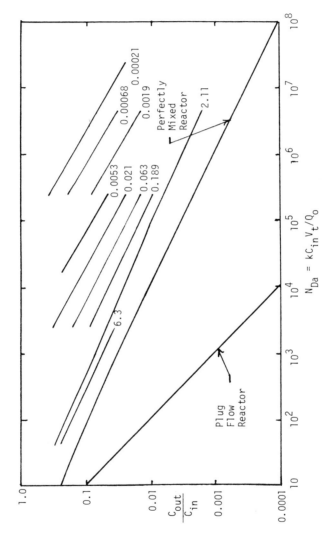

Figure 8   Conversion in the effluent for second-order reactions in a propeller-stirred tank. Propeller centered in tank; $d_T = D/3$; parameters on curves are mixing intensity given by $(\epsilon/L_3^2)^{1/3}_{center}/kC_{in}$.

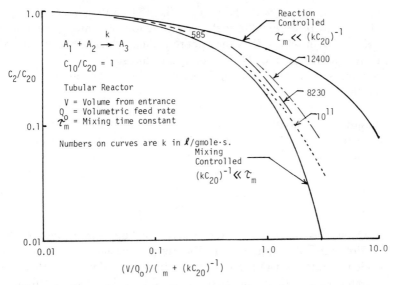

Figure 9   Conversion for second-order reaction in turbulent tubular reactor.

A number of investigators have attempted to apply simple, one-parameter (non-distributed) models of mixing to the complex reaction problem by considering the relative rates of the final mixing process and the chemical reactions. It is generally recognized that complex reactions with at least one fast reaction step are strongly affected by mixing intensity. Cheng and Tookey[24] determined from their data for production of 2,4-dichlorophenoxyacetic acid that the yield correlated best with the stirring power per unit volume. This is roughly equivalent to the turbulence energy dissipation per unit volume. Paul and Treybal[26] correlated the yield of R in the reactions $[A + B \xrightarrow{K_1} R; R + B \xrightarrow{K_2} S]$ as a function of the ratio of a reaction rate term, $K_1\bar{C}_B$, and a mixing rate term, $(1/\tau_M)$, times a ratio of reactant concentrations, giving $K_1\bar{C}_B\tau_M(\bar{C}_{A0}/C_B)$. They evaluated the mixing rate as a function of estimated values of the local turbulence energy dissipation rate per unit mass, ϵ, turbulence intensity, u', and kinematic viscosity, ν, giving $\tau_M = \nu^{3/4}/(\epsilon^{1/4}u')$. These variables were evaluated from previous measurements at the point of feed injection. The correlation was successful for two stirred tank sizes for a given stirrer type, but it showed some uncorrelated effect of reaction rate.

Bourne et al.[27] and Nabholz et al.[28] modelled the relative mixing and reaction rates for a hypothetical spherical eddy of one reactant mixing by diffusion with the surroundings of another reactant. This led them to represent the yield as a function of a mixing modulus $K_2 R^2 \bar{C}_{B0}/D$ in which R is the sphere radius, which is a function of mixing intensity, and D is molecular diffusivity. Nabholz et al. obtained similar behavior for experimental data and model results, but they established no useful relationship between R and the mixing intensity. A similar approach was used by Truong and Methot[29] who assumed a formation of unmixed strands of fluid of the thickness δ. They then modelled the complex reaction as a balance equation for each component in the form as follows:

$$\frac{\partial C_i}{\partial t} = \frac{D\tau}{4\delta^2} \frac{\partial^2 C_i}{\partial y^2} - R_i \tag{25}$$

in which y is the linear distance across a fluid strand and R_i the reaction rate of component i. The term τ designates the reactor reisdence time, and D the reactant diffusivity. They made no effort to connect δ to the actual mixing intensity in the reactor, although the experimental data showed some of the effect on yield given by the model.

Two models which give closely similar results in the one-parameter nondistributed mode are the random coalescence–dispersion (c–d) model[30] and the "Interaction by Exchange with the Mean" (IEM) model[31]. The c–d model may be computed with a Monte-Carlo process or through the use of probability functions[32]; both yield the same results. The reactor is visualized as consisting of many fluid elements or coalescence sites, which may mix (coalesce) in a random fashion with imposed constraints. The number of coalescences per site per element volume, I, was determined by Canon et al.[8] to be as follows:

$$I = 1333(\epsilon/k)(\bar{\tau}/N) \tag{26}$$

where $\bar{\tau}$ is the average residence time in a volume element, N is the number of coalescence sites in the volume element, k is the turbulence kinetic energy per unit mass, and ϵ is the energy dissipation rate per unit mass. If the scale of the fluid elements is small enough, the local mixing process can be successfully simulated. Flow and circulation can also be simulated by moving the sites within the reactor. Simulations of the reactions studied by Paul and Treybal[26] have shown that the method is very successful in predicting mixing effects on selectivity (yield)[8], if the reactor is divided into many small segments (a distributed model), but not if the reactor is considered as a single volume.

The IEM model makes use of the mass transfer equation, $dC_i/dt = h(\bar{C}_i - C_i) - R_i$, in which h is a mass transfer coefficient, and R_i the reaction rate of component i. The term $\bar{C}_i$ is the average (outlet) concentration and C_i the feed concentration to a volume element. In one-parameter models, the time t becomes the fluid element age for use in the residence time distribution. David and Villermaux[31] have found that the usual coalescence dispersion rate, I, equals $4h\tau$ in which τ is the residence time. In a distributed model as developed by Canon et al.[8], τ is the residence time in an individual reactor volume element, see Eq. (26).

None of the non-distributed, one-parameter models can account for the relative effects of bulk circulation, large scale diffusion, and local mixing that lead to spatial distributions of segregation, conversion, and yield in reactors. They also do not include interactions between reactants and turbulent mixing. It is necessary to make use of distributed models and closures with adequate interactions in order to include such effects. Patterson[25] described a model for second-order reactions applied to a segmented reactor that was successful in describing effects of circulation and local mixing in propeller and turbine stirred reactors. By using a finite-difference solution method, Patterson and co-workers applied the same model to a coaxial jet reactor[6]. The model as formulated, however, is not adequate for complex reactions. In the next two sections a c–d model and a closure approximation with interactions between reactants and the turbulent mixing process will be described.

5.2. Coalescence-Dispersion Modelling of Complex Reactions

As mentioned above the c–d model has been shown to do an adequate job in simulating complex reactions in a mixing tank[8]. In the cases simulated the tanks were divided into volume segments each with one-hundred c–d sites. By using the relationship of Eq. (26) to determine the c–d rate in each volume segment, the proper rates of mixing were simulated. Apparently, the scale size and rate of individual mixing occurrences were adequate to, in some way, account for important interactions between reactants in complex reactions and with the physical mixing process. A great advantage of the c–d method is the easy use of any homogeneous reaction kinetics (almost any number of reactions and any order) without any added computational difficulty.

Use of the c–d method becomes particularly complex when large scale turbulent diffusion (such as in jet mixers) is a dominant feature of the mixing process (2-dimensional fluid mechanics). When such turbulent diffusion occurs, not only must c–d sites be moved within the mixer to simulate convection (flow), but coalescence must occur at a frequency and at a distance (to simulate the turbulent mixing scale) to properly model the diffusion. In the tank model of Ref. 7 the sites were moved to simulate the fluid circulation, but coalescences occurred only between adjacent sites. Large scale diffusion was, therefore, not simulated in the tank.

In order to simulate both local mixing and large scale turbulent diffusion, two c–d parameters were made available for adjustment to match the physical process[33]. The c–d rate was again determined by Eq. (26). The diffusion scale was determined by the distance to which coalescence may occur. The model equation used was:

$$L_{c-d} = 60(\nu_e/k^{1/2})(N/V) \qquad (27)$$

where V is the volume of a segment. Computations based on Eq. (27) were made for a complex reaction to determine the yield behavior for a mixing jet, but no comparisons with data have yet been made.

Pratt[34] has provided an alternate treatment of the c–d method for 2-dimensional fluid-mechanics. His approach makes use of a statistical analog to the c–d process so that Monte Carlo computations are not done.

5.3. Proposed Complex Reaction Closure with Interactions

If more than one reaction occurs during the mixing process, closure approximations for finite difference modelling are greatly complicated. One way to simplify the approach to generating such closure approximations is by use of the "paired interaction" assumption. For such multiple reactions the source term in the concentration equation is:

$$P_{\bar{c}_i} = \sum_{k=1}^{m} \sum_{j=1}^{n} \overline{K_{ijk} C_j C_k} \qquad (28)$$

where $m \times n$ = number of reactions, K_{ijk} is positive when $k \neq i$ and $j \neq i$; K_{ijk} is negative when $k = i$ or $j = i$; C_j is one for first-order producing reactions; $K_{ijk} = 0$ for non-reacting combinations. Equation (28) results in a very large number of terms if completely expanded for $C_i = \bar{C}_i + c_i$, where c_i is the fluctuation about the average $\bar{C}_i$. Assuming that K_{ijk} is independent of concentration and that each pair of reactants mixes independently of the presence of the others, in order to avoid many terms, the source term becomes:

$$P_{\bar{c}_i} = \sum_k \sum_j \bar{K}_{ijk} \left\{ \bar{C}_k \bar{C}_j - \sqrt{\overline{c_k^2 c_j^2}}\, (1 - \gamma_{kj})/(1 + \gamma_{kj}) \right\} \qquad (29)$$

where the degree of interdiffusion is

$$\gamma_{kj} = \left[\bar{C}_k \bar{C}_j - \sqrt{\overline{c_k^2 c_j^2}}\, \right] \Big/ \left[\bar{C}_k \bar{C}_j + \sqrt{\overline{c_k^2 c_j^2}}\, \right]$$

A degenerate case of Eq. (29) is where only one second-order reaction occurs, that is, $i = k = 1$ and $j = 2$. Therefore,

$$P_{\bar{C}_1} = \bar{K}_{121} \left\{ \bar{C}_1 \bar{C}_2 - \sqrt{\overline{c_1^2 c_2^2}}(1 - \gamma_{12})/(1 + \gamma_{12}) \right\}$$

which is nearly equal to the result of using Eq. (24) if $\beta (= \bar{C}_{10}/\bar{C}_{20})$ is close to one.

As mentioned following Eq. (24), best results were obtained when the term, $\overline{c_1^2 c_2}$, was set to zero. This is again true when the paired-interaction approach is used. As above, the segregation source term now becomes:

$$P_{c_i^2} = 2 \sum_k \sum_j \bar{K}_{ijk} \left\{ \overline{c_k^2 \bar{C}_j} - \overline{C_k} \sqrt{\overline{c_k^2 c_j^2}}(1 - \gamma_{kj})/(1 + \gamma_{kj}) \right\} - D_{\text{Ti}} + P_{\text{Di}} \qquad (30)$$

where γ_{kj} is defined as above. Again, the reaction term is the same as when Eq. (24) is used if β is close to one.

Eqs (29) and (30) are simplified if the degree of interdiffusion is substituted into them. They become:

$$P_{\bar{C}_i} = \sum_k \sum_j \bar{K}_{ijk} \{ \bar{C}_k \bar{C}_j - \overline{c_k^2 c_j^2}/\overline{C_k}^2 \} \qquad (29a)$$

$$P_{c_i^2} = \sum_k \sum_j \bar{K}_{ijk} \{ \overline{c_k^2 C_j} - \overline{c_k^2 c_j^2}/C_j \} - D_{\text{Ti}} + P_{\text{Di}} \qquad (30a)$$

The terms D_{Ti} and P_{Di} are the same for complex reactions as for single reactions, since they do not depend directly on reaction kinetics. They are as follows:

$$D_{Ti} = 2\overline{c_i^2}/[4.1(k/\epsilon) + (\mu/\rho\epsilon)^{1/2} \ln N_{Sc}] \tag{31}$$

$$P_{Di} = \frac{\partial[\bar{C}_i(C_T - \bar{C}_i)]}{\partial t} + \underline{\nabla} \cdot [\bar{C}_i(C_T - \bar{C}_i)\underline{U}] \tag{32}$$

As explained in Ref. 3, Eq. (31) results from the use of Corrsin's derivation of the rate of segregation decrease in turbulence with a high Schmidt number, N_{Sc}. Eq. (32) allows computation of the segregation increase that would occur, if reaction and turbulent mixing were not present, due to large scale diffusion. In a numerical computation process Eq. (32) is solved simultaneously with the basic transport equations for mass, momentum, turbulence energy, turbulence energy dissipation, concentrations, and segregations. The full set of transport equations for a two-component (single reaction) system was shown in Refs. 5 and 33. The first four equations which model the flow and turbulence were basically the $k - \epsilon$ method of Jones and Launder[12].

5.4. Tests of Closure for Complex Reactions

In order to test the closure relations presented in Eqs. (29) and (30), detailed computations to model typical complex reactions in a turbulent mixing system must be made. Comparisons among these computation results and experimental measurements will then provide a guide for improvement of the closure approximation. Only then may one use such methods with confidence to model a real process, such as combustion or particulate or aerosol formation.

Comparisons of model results have been made with experimental data obtained in agitated tank reactions with kinetics of the type

$$A + B \xrightarrow{k_1} R; R + B \xrightarrow{k_2} S.$$

Such comparisons were made possible through use of the hydrodynamic mixing model for stirred tanks described in Ref. 25, in which the flow and turbulence are not modelled but are scaled from experimental data. No comparisons with experimental data from shear flow mixing systems (jets or mixing layers) have been made, because the proper data have not yet been found. In order to make a good comparison, the flow system must be simple and well described, and the reactions must be relatively simple and few in number, so that the modelling method is not overwhelmed.

5.5. Stirred-Tank Continuous Flow Reaction

Truong and Methot[29] measured the yield of product R in a reaction scheme where $k_1/k_2 = 2$ and $C_{A_0}/C_{B_0} = 1$. The Damköhler number, $N_{Da} = k_1 C_{A_0} \bar{\tau}$, for the stirred tank was in the range 25–150.

The hydrodynamic tank mixing model mentioned above was applied to model the Truong and Methot experiments. The tank was divided into 30 mixing segments in order to model the distribution of flow, turbulence, concentration and segregation of each component. The modelling method is equivalent to finite difference modeling, where velocities and turbulence energy dissipation rates and scales are specified.

Figure 10 shows a comparison of a wide range of results obtained from the model with the limited experimental data of Truong and Methot. The results are presented in terms of the conversion of component B which reacts $(\bar{C}_{B_0} - \bar{C}_{B_{final}})$ to form product R as a function of Damköhler number, N_{Da}. The plotting parameter which naturally resulted from the model results is a measure of the ratio of impeller stream flow rate to reactant feed rate, ND^3/Q_0. In terms of a shear flow mixing system, that is equivalent to lateral diffusion rate to flow rate.

Although much more experimental data are necessary to confirm the results of the model for stirred tanks, Fig. 10 seems to indicate that the model approaches the real behavior.

6. CONCLUSION

Scaleup of turbulently mixed reactors is not a simple and straight-forward problem. In most instances a full modelling approach must be used to compare results at different sizes. The good-old scaling rules for stirred tanks, for instance, based on power dissipation per unit volume to some exponent have some basis since local energy dissipation per unit volume is involved in most mixing rate models, but the complex interactions of convection, turbulent diffusion, mixing rate, and chemistry make scaleup a many parameter problem. In order to improve our ability to predict accurately the

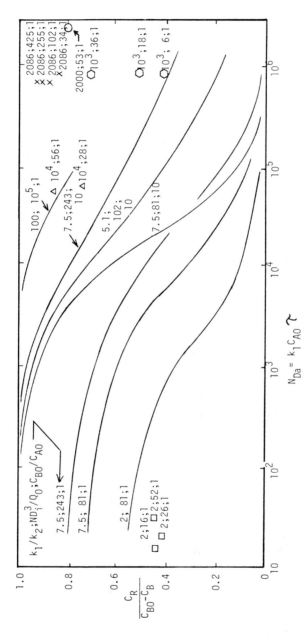

Figure 10   Prediction of the yield of R in the reactions $A + B \xrightarrow{k_1} R$, $R + B \xrightarrow{k_2} S$ in a Rushton turbine stirred tank. The "Paired-Interaction" version of the HDM model was used.   Truong and Methot[29]

behavior of large reactors involving turbulent mixing, fundamental methods of modelling are a necessity.

7. NOTATION

A_i chemical component i
C_i concentration of component i
c_i concentration fluctuation
C_1, C_2 constants in Eq. (7)
Cg_1 constant in Eq. (14)
Cg_2 constant
C_{EBU} eddy breakup constant
D_i diffusivity of component i in a mixture
D diameter
g body force
H enthalpy of fluid
h enthalpy fluctuation or mass transfer coefficient
I coalescence–dispersion rate
K_{ijk} reaction rate constant for reaction between components i and j to produce component k
k_{ijk} reaction rate constant fluctuation
k turbulence energy, $\overline{u_i^2}/2$
l length scale of turbulence
L_{c-d} scale of coalescence–dispersion diffusion
n number of moles
N_{Sc} Schmidt number
N_{Da} Damkohler number
N number of coalescence sites
p pressure
$P_{\bar{C}_i}, P_{\overline{c_i^2}}$ production terms
Q_0 volumetric feed rate
R_i rate of production of component i
R radius of hypothetical spherical eddy
S source term in a balance equation
t time
T temperature
U_i fluid velocity in i direction
u_i velocity fluctuation
V volume of a reactor segment
x_i coordinate distance in i direction
y distance across a fluid strand

Greek Symbols

β reactant ratio

γ extent of interdiffusion

δ thickness of fluid strand

ϵ rate of turbulence energy dissipation per unit mass

ν kinematic viscosity

ρ fluid density

σ ratio of total diffusivity to turbulent plus molecular kinematic viscosity

τ time constant or residence time

ξ mixture fraction

Subscripts

i, j, k, l, n coordinate direction or chemical component

$+, -$ upper or lower extents of fluctuation

e pertaining to a turbulent quantity

m, M due to mixing

k pertaining to turbulence energy

$\bar{c}$ pertaining to concentration

c^2 pertaining to segregation

r, R due to reaction

s stoichiometric

T total or tank

ref   reference level

0 initial or feed

underbar   vector

REFERENCES

1. Levenspiel, O. (1962). *Chemical Reaction Engineering*, Ch. 9, Wiley, New York.
2. Nauman, E. B. (1981). *Chem. Eng. Comm.*, **8**.
3. Corrsin, S., *AIChE J.*, **10**, 870 (1964).
4. Patterson, G. K. (1973). Fluid mech. of mixing, *Proc. of A.S.M.E. Meeting*, Fluids Eng. Div., Atlanta.
5. Spalding, D. B. (1971). *Chem. Eng. Sci.*, **26**, 95.
6. Patterson, G. K., Lee, W.-C., and Calvin, S. J. (1977). *Pres. at Symp. on Turbulent Shear Flows*, Penn. State University.
7. Rao, D. P., and Dunn, I. J. (1970). *Chem. Eng. Sci.*, **25**, 1275.
8. Canon, R. M., Wall, K. W., Smith, A. W., and Patterson, G. K. (1977). *Chem. Eng. Sci.*, **32**, 1349.
9. McKelvey, K. N., Yieh, H.-N., Zakanycz, S., and Brodkey, R. S. (1975). *AIChE J.*, **21**, 1165.
10. Elghobashi, S. E., Pun, W. M., and Spalding, D. B. (1977). *Chem. Eng. Sci.*, **32**, 161.

11. Lockwood, F. C., and Naquib, A. S. (1975). *Combustion and Flame*, **24**, 109.
12. Jones, W. P., and Launder, B. E. (1973). *Int. J. Heat Mass Trans.*, **16**, 119.
13. Launder, B. E., and Spalding, D. B. (1972). *Mathematical Models of Turbulence*, Academic Press, New York.
14. Hill, J. C. (1979). *Chem. Eng. Educ.*, Winter, p. 34.
15. Brodkey, R. S. (1967). *Phenomena of Fluid Motions*, Addison-Wesley, Reading, Massachusetts.
16. Bilger, A. (1979). *Turbulent Reacting Flows*, (Eds. P. A. Libby and F. A. Williams), Springer-Verlag, Heidelberg, Chap 3.
17. Gosman, A. D., Run, W. M., Runchal, A. K., Spalding, D. B., and Wolfshtein, M. (1969). *Heat and Mass Transfer in Recirculating Flows*, Academic Press, New York.
18. Spalding, D. B. (1978). *Turbulent Combustion*. (Ed. L. A. Kennedy), AIAA, p. 105.
19. Borghi, R. (1975). *Turbulent Mixing in Nonreactive and Reactive Flows* (Ed. S. N. B. Murthy), Plenum Press, New York, p. 163.
20. Donaldson, C. duP. (1975). *Turbulent Mixing in Nonreactive and Reactive Flows* (Ed. S. N. B. Murthy), Plenum Press, New York, p. 131.
21. Ellail, M. M. M. A., Gosman, A. D., Lockwood, F. C., and Megahed, I. E. A. (1978). *Turbulent Combustion* (Ed. L. A. Kennedy), AIAA, p. 163.
22. Patterson, G. K. (1979). *Proc. 2nd Symp. on Turbulent Shear Flow*, Imperial College, London.
23. Patterson, G. K. (1974). *Proc. 1st Eur. Conf. on Mixing and Centrif. Sep.*, Paper A4, BHRA Fluid Eng., Cambridge, England.
24. Cheng, D. C. H., and Tookey, D. J. (1977). *2nd European Conf. on Mixing*, Cambridge, England.
25. Patterson, G. K. (1975). *Application of Turbulence Theory to Mixing Operations* (Ed. R. S. Brodkey), Academy Press, New York.
26. Paul, E. L., and Treybal, R. E. (1971). *AIChE J.*, **17**, 718.
27. Bourne, J. R., Moergeli, U., and Rys, P. (1977). *2nd European Conf. on Mixing*, Cambridge, England.
28. Nabholz, F., Ott, R. J., and Rys. P. (1977). *2nd European Conf. on Mixing*, Cambridge, England.
29. Truong, K. T., and Methot, J. C. (1976). *Can. J. Chem. Eng.* **54**, 572.
30. Rao, D. P., and Rao, A. R. (1974). *Chem. Eng. Sci.*, **29**, 1809.
31. David, R., and Villermaux, J. (1975). *Chem. Eng. Sci.* **30**, 1309.
32. Evangelista, J. J. (1970). Ph.D. Thesis, City College of New York.
33. Patterson, G. K. (1977). Modeling complex chemical reactions in flows with turbulent diffusive mixing, *70th Ann. AIChE Meeting*, New York.
34. Pratt, D. T. (1976). *Prog. Energy Combus. Sci.* **1**, 73.
35. Vassilatos, G. and Toor, H. L. (1965). *AIChE J.* **11**, 666.

Chapter 4

Mixing of Viscous Non-Newtonian Liquids

JAROMIR J. ULBRECHT

Department of Chemical Engineering, State University of New York, NY 14260, USA, †

and

PIERRE CARREAU

Department of Chemical Engineering, Ecole Polytechnique of Montreal, Quebec, Canada H3C 3A7

1. INTRODUCTION

In many industrial mixing operations, the liquid raw materials, products, and intermediates (or all three of them) are rheologically complex. Numerous examples can be found in the polymer based industries (manufacturing and processing of synthetic rubbers, plastics, fibers, resins, paints, coatings, and adhesives), food-processing industries, bio-chemical operations, and in the manufacturing of detergents, fertilizers, explosives, and propellants. The spectrum of rheological complexities is wide: paints are thixotropic and dispersions may have a yield stress. The most frequently occurring anomalies are, however, a shear-dependent viscosity and visco-elasticity. Both these phenomena are characteristic of fermentation broths and of polymer solutions and melts though, by no means, exclusive to them only.

There are two fairly distinct regions in the polymer based industries where mixing of non-Newtonian liquids will be met:

†Present address: Center for Chemical Engineering, National Bureau of Standards, Washington, D.C. 20234, USA

polymer manufacturing and polymer processing. In the latter, polymers are mixed as they pass through extruders, calenders, and other processing equipment. It is fair to say that no effort has been spared by the industry to analyze these processes and to develop sophisticated models which would predict the flow field and thus contribute to the development of more objective design methods. The manufacturing section of the polymer industry has not received, so far, the same attention and yet, some of the ultimate polymer properties like the average molecular weight, the molecular weight distribution, the extent of cross-linking, and others are fixed in the polymerization reactor (they can, of course, deteriorate during the subsequent processing). About two thirds of the world production of rubbers, plastics, and fibers are made by solvent or bulk polymerization where both the rate and the quality of mixing in the polymerization reactor will control, to a large extent, the final properties of the product.

The advent of biochemical reactors has focused the attention of engineers–rheologists on the flow behaviour of fermentation broths which may display a perplexing variety of rheological complexities. Because of the dispersion of cellular biomass in a substrate rich in polymer, which may be either added to the batch or produced by the reaction, the fermentation broth is not only shear-thinning but it will show some degree of elasticity and a yield stress as well[1].

The pioneering work of Metzner[2,3] on the implications of shear-dependent viscosity on the flow in stirred tanks has already led to the establishment of well tested design procedures but the concept of viscoelasticity has not become yet a household term everywhere. Although the implications of the viscoelastic behaviour on the velocity pattern, mixing and circulation times, and power consumption in stirred tanks were analyzed in the literature some time ago[4,5], the subject is still considered esoteric rather than real.

Two excellent reviews of mixing of non-Newtonian liquids were published recently: A design-oriented one by Skelland[6] and an earlier one by Chavan and Mashelkar[7] placing more emphasis on research techniques. The rheology of polymers is a complex and rapidly growing area of research and the reader is referred to the texts of Bird et al.[8] and of Middleman[9].

2. BASIC CONSIDERATIONS

2.1. Flow Simulation

The influence of the two rheological anomalies, the shear-dependent viscosity and the viscoelasticity, can be demonstrated using an experiment in which the impeller is simulated by a coaxial rotating cylinder (Figs. 1 and 3). Parts of the following treatment have been taken from the review papers cited above[4,5].

If the liquid in the tank is a Newtonian one and the flow is laminar, then the angular velocity distribution is a quadratic function of the radial distance

$$(\omega/\Omega)_N = (1 - S^2)^{-1}[1 - S^2(R_1^2/r^2)] \qquad (1)$$

where S is the ratio of the vessel to the impeller diameter. If this quadratic velocity profile is used to calculate the relationship between the torque per unit length of the cylinder and its angular velocity Ω then this relationship is found to be linear:

$$M(S^2 - 1)/2\pi R_1^2 S^2 = 2\mu\Omega \qquad (2)$$

with the coefficient of proportionality μ known as the viscosity of the liquid. The validity of Eq. (2) has been confirmed experimentally many times over.

If the same experiment is repeated with a polymer melt or solution, with a fermentation broth, or with a detergent solution then, as shown in Fig. 1, the angular velocity is found to drop much

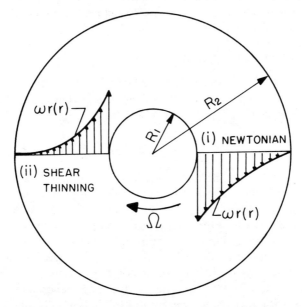

Figure 1 Velocity distribution in the primary flow (the action of the mixer is simulated by that of a coaxial cylinder of radius R_1). (i) Newtonian liquid, (ii) shear-thinning liquid.

faster

$$(\omega/\Omega)_{nN} < (\omega/\Omega)_N$$

than expected on the basis of Eq. (1). This phenomenon was for the first time analyzed by Metzner et al.[2,3] in vessels stirred by turbines. As a consequence of this behaviour, the ratio of the torque M to the angular velocity Ω is no longer a constant for these liquids and the viscosity in the sense of Newton's law of viscosity

$$\tau_{21} = -\mu\dot{\gamma} \tag{3}$$

has no physical meaning. Since we are, however, used to the concept of viscosity we shall continue using the ratio

$$M(S^2 - 1)/4\pi\Omega R_1^2 S^2 = \eta_e \tag{4}$$

as a measure of the viscous property of the liquid, calling it the "effective viscosity". Our experiment will soon reveal that the effective viscosity is not a material property but that it varies with both the geometry of the mixer as well as with the rotational speed of the impeller. Since both these variables determine the shear rate in the mixer, we recognize the effective viscosity as being shear-dependent.

2.2. Non-Newtonian viscosity

We make a distinction between the effective viscosity and the non-Newtonian viscosity, which is defined by extension of Newton's law of viscosity for simple shear flows[8] as

$$\eta \equiv -\tau_{21}/\dot{\gamma} \tag{5}$$

The non-Newtonian viscosity is a material function which is independent of the flow geometry. It depends, however, on the temperature (and eventually on the pressure) and on the rate of shear or shear stress in the liquid. In a complex flow such as the flow generated by an agitator, the non-Newtonian viscosity is a local property which varies with the position in the vessel whereas the effective viscosity is an overall average property. For the vast majority of non-Newtonian liquids, the non-Newtonian viscosity decreases with increasing shear rate so that we classify them as "shear-thinning". Some typical viscosity curves are shown in Fig. 2

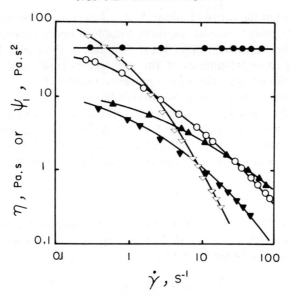

Figure 2   Variation of the viscosity (solid points) and of the primary normal stress coefficient (plain points) with the rate of shear. ●○ polysiloxane (silicone oil); ▼▽ polyacrylamide solution; ▲ CMC solution (normal stress negligible).

for aqueous solutions of two water-soluble polymers and a medium viscosity polysiloxane (silicone oil).

Without any claim to rigour or generality it is convenient to describe the viscosity curves by empirical formulae, commonly known as rheological models. The simplest and most frequently used model is the "power-law"

$$\eta = m|\dot{\gamma}|^{n-1} \qquad (6)$$

where the consistency m and the flow index n are material parameters both being functions of temperature only. For shear-thinning non-Newtonian liquids, the flow index n has a value smaller than one. On the other hand, n is larger than one for shear-thickening liquids. When the shear rate approaches zero the power-law predicts for shear-thinning a viscosity which is too high. Usually, the power-law expression will not be valid over more than two decades of shear rate and any extrapolation beyond these limits could lead to large errors. This should be kept in mind, mainly in designing large scale mixing equipments using data obtained from a laboratory scale system. The data for the small scale equipment may well be within the range of applicability of the power-law expression but a large mixer would normally rotate at much smaller speed, hence at low effective rate of shear. The use of the power-law may then lead to overestimates of the effective viscosity and to large errors in the design value of the power requirement.

There are, however, non-Newtonian liquids, such as pastes, slurries, and certain polymer solutions (Carbopol) the viscosity of which rises with decreasing shear rate faster than the power-law model can predict. In such a case, another term can be added to Eq. (6) and the resulting model is known as the Herschel–Bulkley equation

$$\eta = \tau_y/|\dot{\gamma}| + m|\dot{\gamma}|^{n-1} \qquad (7)$$

where τ_y is termed the "yield stress". A two-parameter version of Eq. (7), in which n is put equal to one and $m = \mu_p$, is known under the name of the Bingham model

$$\eta = \tau_y/|\dot{\gamma}| + \mu_p \qquad (8)$$

Most polymeric materials will show a finite viscosity when the shear rate approaches zero rather than an infinitely high viscosity. The viscosity of these materials can be well modelled by a Carreau formula

$$\frac{\eta - \eta_\infty}{\eta_0 - \eta_\infty} = [1 + (\lambda_c\dot{\gamma})^2]^{(n-1)/2} \qquad (9)$$

which interprets two plateaus, a "zero-shear viscosity" η_0 and a "high-shear limiting viscosity" η_∞. The flow index n has the same meaning as that in the power-law expression and the parameter λ_c is a liquid's characteristic time. The high-shear limiting viscosity is usually very small compared to the zero-shear viscosity and for most engineering applications it can be neglected. The Carreau model is then quite similar to another popular model, the Ellis model

$$\eta/\eta_0 = [1 + (\tau_{21}/\tau_{1/2})^{\alpha-1}]^{-1} \qquad (10)$$

where α is related to the slope of the viscosity curve in the power-law region and $\tau_{1/2}$ is the shear stress at which the viscosity is equal to one half of the zero-shear viscosity η_0. The combination of Ellis parameters $(\eta_0/\tau_{1/2})^\alpha$ can be compared with Carreau's λ_c. Both the Carreau and Ellis models have been shown to describe very well the viscosity data of various polymer solutions[10]. Another version of a three-parameter model is the Cross–Williamson model

$$\eta/\eta_0 = [1 + |\lambda_c \dot{\gamma}|^{1-n}]^{-1} \tag{11}$$

2.3. Liquid's Elasticity

The shear-dependent viscosity is, however, only part of the non-Newtonian story. Everyone who has stirred a thick polymer solution has noticed how a part of the liquid climbed up the shaft of the rotating impeller. This peculiar behaviour contradicts all experience with ordinary liquids when the liquid surface is expected to rise along the wall of the tank due to the centrifugal forces (Fig. 3). It is thus obvious from this experiment that there is some new force generated in the liquid by the rotational flow which competes with the centrifugal force of inertia and which drives the liquid towards the shaft of the impeller. This force cannot be associated either with viscous or inertial forces (even less with gravity) and it is normally attributed to the liquid's elasticity.

Intuitively one associates elasticity with some sort of rubber-like behaviour, that is, with the tendency to regain original shape after deformation. Let us imagine for a while one streamline or a liquid thread in our mixing thought-experiment. If one admits that the liquid has, apart from viscous, also elastic properties then part of the energy transported from the impeller will not be dissipated by viscous friction but it will be stored in the liquid by stretching polymer molecules like a rubber string. Due to such deformation, a

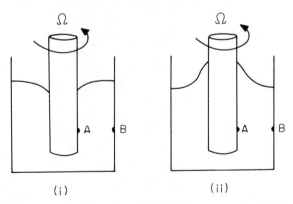

Figure 3   Shape of the liquid's free surface in a (i) Newtonian liquid, (ii), viscoelastic liquid. In the Newtonian liquid, the pressure reading by a transducer at point A is smaller than the pressure reading at B. In the viscoelastic liquid, $P_A > P_B$.

tension develops in the liquid elements compelling them to relax the tension by shrinking to a shorter length. This is achieved by assuming a shorter radial distance from the center of rotation. The result of this is a net force normal to the shear plane acting towards the axis of rotation. It is just like winding a rubber string around a finger: the more you stretch it the more you feel the normal force on your finger.

In order to interpret this elastic force in the sheared liquid let us make use again of the vessel stirred by a coaxial cylinder (Fig. 4). A liquid element is marked between the vessel wall and the impeller which is sufficiently small so that no liquid properties change across the element. Because the liquid is stationary at the vessel wall but it rotates with the cylindrical impeller, a velocity gradient between these two surfaces exists which gives rise to a shear force. This is shown in Fig. 4 in terms of the shear stress τ_{21}. (In our idealized experiment, there are no velocity gradients in z and θ directions so that the other two shear stress components are equal to zero.) It has been already shown that, for non-Newtonian liquids, the relation-

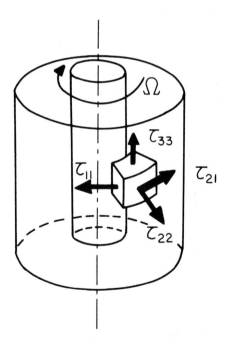

Figure 4 State of stresses in a vessel stirred by a cylinder; τ_{11}, τ_{22}, τ_{33} are normal stress components, τ_{21} is the shear stress component.

ship between τ_{21} and $\dot{\gamma}$ is a non-linear one which results in a velocity profile significantly different from that of a Newtonian liquid. The shear stress τ_{21} is, however, not the only stress component acting on our element of fluid. There are three normal stress components, τ_{11}, τ_{22}, τ_{33}, acting perpendicularly on the faces of the element. In a Newtonian liquid, all these normal stresses are equal to zero irrespective of whether the liquid flows or is at rest. Not so in a viscoelastic liquid in which each of these stresses is a function of the velocity gradient. For incompressible fluids, only two of the normal stresses are independent and two additional material functions are defined (under steady simple shear flow)

$$\psi_1 \equiv -(\tau_{11} - \tau_{22})/\dot{\gamma}^2 = N_1/\dot{\gamma}^2 \tag{12}$$

and

$$\psi_2 \equiv -(\tau_{22} - \tau_{33})/\dot{\gamma}^2 = N_2/\dot{\gamma}^2 \tag{13}$$

The term ψ_1 is called the primary normal stress coefficient. As shown in Fig. 2, the primary normal stress coefficient is of the same order of magnitude as the non-Newtonian viscosity of the liquid and the respective curves have similar shapes (with the exception of the silicon oil which is viscoelastic but has a constant, that is, shear-independent viscosity). At large enough shear rates, the primary normal stress difference, N_1 can be much larger than the shear stress, τ_{21}. Very few data of the secondary normal stress coefficient, ψ_2 are available. Serious difficulties are encountered in the measurement of the secondary normal stress difference, N_2. It should be noted, however, that ψ_2 is much smaller than ψ_1 (about 10% of ψ_1) and is negative. Therefore it is believed that ψ_2 has little influence on the hydrodynamics in the mixing vessel and it will be ignored in the rest of this text.

Like the viscosity curve, the primary normal stress curve can be described by an empirical interpolation formula

$$\psi_1 = m'|\dot{\gamma}|^{n'-2} \tag{14}$$

where m' and n' are material parameters being functions of temperature only. It must be kept in mind that equations like (6) or (14) have no theoretical foundation and that they can be used only as interpolation formulae within the range of shear rates in which they were measured. In this context it must be also remembered that most rheological measurements will yield values of rheological parameters obtained in one-dimensional steady-shear flows. It is, by no means, certain that we may use these for such complex three-dimensional flows as they exist in a stirred tank. At present, however, no better technique is available. This is not because we would lack rigorous rheological equations of state predicting fairly accurately the spatial state of stresses in a flowing liquid, but because nobody has attempted yet to solve (numerically) the three dimensional equations of motion, coupled with those equations of state.

Being now aware of the existence of non-zero normal stress differences in flowing non-Newtonian liquids we can explain the elastic force competing with inertia. It follows from the equation of motion that the pressure difference between the inner and the outer cylindrical walls of our thought experiment, at a given horizontal plane, is equal to

$$\Delta P = P_A - P_B = -\rho \int_{R_1}^{R_2} r\omega^2 \, dr + \int_{R_1}^{R_2} [N_1/r] \, dr \qquad (15)$$

It will be immediately recognized that the first term in Eq. (15) is the contribution of inertia (centrifugal force), the second is that of the extra normal stresses. Because for Newtonian liquids, $N_1 = 0$, the pressure difference has a positive value depending only on the density of the liquid ρ and the velocity $r\omega$. In a viscoelastic liquid, however, the primary normal stress difference, N_1 is no longer equal to zero. It has its lowest value near the vessel wall and its highest value near the rotating impeller. Thus, the second term in Eq. (15) has an opposite sign than the inertia term, in other words the elastic term opposes the centrifugal force. If the elastic force is sufficiently high it will overcome the inertia and the liquid will be thrown towards the impeller (Fig. 5).

The generation of non-zero normal stress differences is not the only manifestation of viscoelasticity in non-Newtonian liquids but it is the most significant manifestation in steady-state flows. The flow around an impeller has, however, an element of periodicity since the impeller consists of a finite number of blades or paddles passing periodically through the liquid. In the close vicinity of the impeller, the analogy with a rotating cylinder will fail and another propensity of viscoelasticity will come to play a role: the relaxation time, λ. In simplest terms, the unsteady response of a viscoelastic liquid can be

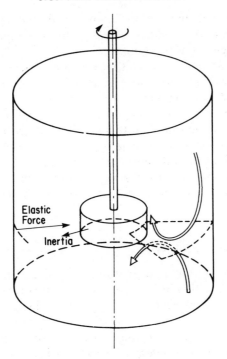

Figure 5   Forces and streamlines in a stirred tank.

pictured by a Maxwell model

$$\tau_{21} + \lambda\,(\partial\tau_{21}/\partial t) = -\mu\dot{\gamma} \qquad (16)$$

in which the relaxation time, λ is related to the elastic modulus G by $\lambda = \mu/G$. If the relaxation time is of the same order of magnitude as the reciprocal of the characteristic frequency of the impeller then it may play an important role in the hydrodynamics of a mixing system.

Another manifestation of viscoelasticity is the reduction of turbulent drag known also as the Tom's effect. As the term suggests, this phenomenon occurs only under turbulent flow conditions and it is likely to be encountered only in very dilute solutions of long chain molecules the viscosity of which is close to that of the solvent. It may also come up in those biochemical reactors in which the viscosity is controlled by additives with elastic properties[1]. Although this phenomenon is not yet well understood it is fair to say that the drag reducing polymers interfere only with the low end of the spectrum of turbulent eddies. Some of the consequences are discussed in Chapter 8 of this book.

The dramatic increase of the extensional viscosity of some polymer melts or solutions with increasing extensional rate is also often identified as being one of the manifestations of viscoelasticity. This phenomenon plays an important role in the processes of gas and liquid dispersion since before a bubble or drop can break down, it must be exposed to a fast extensional flow. Finally an increasing extensional viscosity will also affect the rate of coalescence of bubbles and drops since it will slow down the drainage of the liquid from the film which separates the drops or bubbles after they collided.

2.4. Dimensionless Numbers

The complexity of flow in mixers of any type puts the dimensional analysis in the position of being an exclusive theoretical tool for the correlation of experimental data. When working with non-Newtonian liquids we must be prepared to put up more dimensionless groups than we are used to since each new rheological parameter justifies a new dimensionless number. Further, since there is no constant viscosity, the dimensional analysis will yield a Reynolds number utilizing any parameter having the dimension of viscosity (such as the zero-shear viscosity). The flow index of the power-law model, being itself dimensionless, rarely figures as a dimensionless number on its own. More frequently, it is incorporated in the Reynolds number in one way or another but always respecting the requirement that the non-Newtonian Reynolds number reduces to its Newtonian form for $n = 1$. When working with the Ellis model, we shall have the Ellis number defined as the ratio of the characteristic time of the liquid ($\eta_0/\tau_{1/2}$) to the characteristic time of the process. Obviously the choice of the characteristic process time will depend on the nature of the process and we can only advise the reader to use his or her sound judgement. The Ellis model is not unique in providing a combination of material parameters having the dimension of time: an analogy to the Ellis number can be formed using the Carreau and other models. These are summarily known as the "viscosity numbers". A thorough analysis of this subject was recently presented by Skelland[6] and it will not be repeated here. We shall, however, say a few more words about the dimensionless numbers characterizing the viscoelastic behaviour. It has been already said that one measure of viscoelasticity is the primary normal stress coefficient ψ_1. The ratio of this coefficient to the viscosity multiplied by a characteristic rate of the process is known as the Weissenberg number

$$\text{Wi} = (\psi_1/\eta)\Omega \tag{17}$$

(the symbol Wi is preferred before We to avoid confusion with the Weber number) which measures the extent of elastic forces in terms of viscous forces. The Weissenberg number is related to, but should not be confused with the Deborah number

$$\text{De} = \lambda\Omega \tag{18}$$

which relates the elastic relaxation time of the liquid to the characteristic time of the process. While the Weissenberg number will play a controlling role in steady-state flows, the Deborah number will enter the correlations whenever a transient or a periodic process occurs. The measurement of the relaxation time, however, presents many difficulties and λ is usually obtained from steady-shear data through the use of an appropriate constitutive relation. For example, the convected Maxwell model proposed by White and Metzner[11] gives λ equal $\psi_1/2\eta$. In such a case the Weissenberg and Deborah numbers, defined here for mechanically agitated systems, convey the same information.

Finally, it needs to be stressed again that a judicious choice of the characteristic process time is crucially important for any correlation to be successful. Other choices than Ω may be in order when dealing with two-phase systems.

3. FLOW PATTERN IN STIRRED TANKS

The pattern of flow in a stirred tank is the result of the geometrical configuration of the mixer, of the properties of the liquid, and, to some extent, of the rotational speed of the agitator. The flow pattern in the stirred tank is solely responsible for the quality of mixing, both on the macro- as well on the microscale. On the macroscale, an unfavourable flow pattern will result in the formation of pockets of stagnant liquid and of short-cuts which, in turn, will lead to a poor utilization of the mixing volume and to long tails of the residence time distribution curves. On the microscale, a flow pattern which lacks sufficient high stretch and shear rates will fail to create adequate opportunity for mass transport.

3.1. Main Flow Pattern

It is convenient to classify the agitators used in the many non-Newtonian technologies into three classes: (i) Those which operate at high speeds, creating thus locally high shear rates and relying on good momentum transport to carry the energy from the impeller into the far corners of the vessel; a typical example is a turbine impeller. (ii) Those which cannot rely on adequate momentum transport because of viscous damping and which, therefore, are large in order to reach into the far corners of the vessel. Examples include gate and anchor impellers. (iii) Finally, we have those which operate slowly without creating high gradients but which rely on their excellent pumping capacity to reach every corner of the vessel; these are the helical screw and helical ribbon agitators.

Irrespective of whether the impeller falls into the first, second, or third class, the primary flow pattern in the liquid is a tangential one as the liquid follows the impeller rotating about its own axis. Helical agitators, pitched blade turbines, and propellers generate also an axial component of flow so that the primary flow pattern is helical. Because of the centrifugal force generated by the primary tangential flow, a secondary radial flow is created heading away from the impeller towards the walls of the container. There, the secondary flow becomes vertical only to become radially inward at the bottom and at the top of the tank. The competition of this inertia driven secondary flow with the flow driven by elastic forces has been already mentioned.

The primary and the secondary flow patterns are common to all rotational flows, be it around impellers or around simpler bodies, such as rotating spheres, disks, and other. This fact has led to the simulation of flows around impellers and the results of this simulation will be mentioned later in this section.

There is, however, a tertiary flow pattern peculiar to rotating agitators and totally absent from the flows around spinning disks and spheres: the stagnation flow on the upstream face of a blade leading to the vortex flow on the downstream faces of the agitator blades. In the following paragraphs, we shall deal briefly with these three flow patterns and how they are influenced by the rheological complexities of the liquid.

3.2. Non-Newtonian Effects

The influence of the shear-dependent viscosity on the velocity pattern in a stirred tank is relatively easy to predict: the class (ii) and (iii) impellers are going to perform better under these circumstances but the effect will not be dramatic unless the apparent viscosity increases steeply with falling shear rate.

Metzner and Taylor[3] measured the primary (tangential) velocities for turbine impellers (class (i)) operating in a number of shear-thinning liquids and found the tangential velocities decreasing much faster with increasing radial position than they did in liquids with a constant (shear-independent) viscosity. In severe cases, such as are the liquids with yield stress, Wichterle et al.[12] and Nienow et al.[13] found the impeller cutting its way in a small fluid cavern beyond which the liquid was totally stagnant. Obviously, impellers which do not rely on the viscous transport of momentum are free of this problem although Peters and Smith[14] found a considerable damping of the tangential velocity profile when using an anchor agitator. If there is insufficient tangential flow, the inertial force is too weak to generate a secondary flow and the quality of mixing suffers.

The influence of viscoelasticity on the flow pattern is much more striking, though more difficult to assess. This is partly because modelling of three-dimensional viscoelastic flows is still in its infancy, partly because most viscoelastic liquids are also shear-thinning so that separating these two phenomena experimentally is difficult. An attempt to solve the governing equations of motion would meet with enormous difficulties. The complexities are essentially due to the complicated shape of the agitator, the non-linear terms in the rheological equation of state, as well as the non-linear inertial terms in the equation of motion notwithstanding the fact that a numerical solution would be of very limited utility in that it would be linked to one particular set of boundary conditions.

The problem could be somewhat simplified by removing the complexities due to the shape of the impeller and replacing it with simple geometrical bodies such as spheres or disks. The solution of the equations of motion can be obtained in these cases with greater ease and the kinematics and the dynamics of flow around these bodies can be described in more quantitative terms even though the simplifications accepted during such an analysis may limit the extent of the agreement between the predictions of the model and the observed patterns. It must be borne in mind, however, that only the primary and the secondary flow patterns can be simulated in this way while the vortex flow behind the blades remains ignored.

Perhaps the first photographs of the secondary flow pattern (made visible by an injection of a dye) around a sphere, a turbine and a propeller rotating in a viscoelastic liquid were presented by Giesekus[35]. In all the three cases the impeller was surrounded by a small region in which the liquid moved centrifugally out of the impeller region but in the rest of the vessel the flow pattern was opposite with the liquid flowing towards the impeller in the equatorial plane and outwards from the impeller along the axis of rotation. A closed streamline separates these two regions so that there is no convective mass transport between them, equally as there is no convection between the portions of the liquid above and below the equatorial plane. More photographic evidence of this complex secondary flow pattern can be found in the work of Feldkamp[15] and Uemura et al.[16] The primary and secondary velocity vector field measured by a particle tracer technique was obtained by Kelkar et al.[18] One of the important contributions of this work is the observation that, whatever the secondary flow pattern may be, the primary (tangential) flow around the rotating body is not affected by the extent of viscoelasticity of the liquid as long as the Reynolds number is less than 5. This finding will help us understand why the liquid's elasticity has so little influence on the torque in the viscous regime of flow.

In an attempt to quantify this information, we can turn to the work of Walters and Savins[19] who predicted three modes of the secondary flow pattern around a sphere rotating in a viscoelastic liquid, depending on the magnitude of the ratio Wi/Re. If this ratio is smaller than 1/6 then inertia dominates the whole flow field and the flow pattern is as shown in Fig. 6a. If, on the other hand, this ratio is larger than 1/2, then the elastic forces dominate the whole flow field and the flow pattern is described by Fig. 6b. For the intermediate region the flow pattern is depicted in Fig. 6c. How do these predictions compare with the experimental evidence? Although the existence of all these three modes has been experimentally confirmed, the occurrence of the Fig. 6c pattern is rather unlikely. The most frequently observed pattern around rotating spheres, disks, turbines, and propellers is that shown in Fig. 6d which was not predicted by the theory of Walters and Savins[19]. Another, rather serious deviation is the strong dependence of the flow pattern on the rotational speed of the impeller which is not predicted by the theory (the reader will notice that the ratio Wi/Re for a third-order liquid does not contain the velocity of the impeller

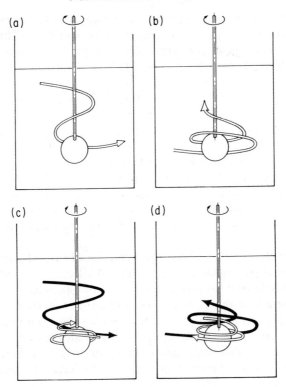

Figure 6   Instability of the flow pattern. (a) Flow dominated by inertial force; (b) Flow dominated by elastic force; (c) Nearside flow (unshaded streamline) dominated by elasticity, offside flow (shaded streamline) dominated by inertia; (d) Nearside flow (unshaded streamline) dominated by inertia, offside (shade streamline) by elasticity.

and the viscosity of the liquid). The explanation of these discrepancies can be found in the properties of the rheological equation of state used by Walters and Savins[19] which yields a constant (shear-independent) viscosity and a constant primary normal stress coefficient. An inspection of Fig. 2 will show that these are rather unrealistic assumptions. Near the sphere, where the shear rate is highest, the primary normal stress coefficient attains its minimum value and, thus, the ratio Wi/Re may drop below its critical magnitude allowing the inertial forces to dominate the region. Despite these discrepancies, the theory of Walters and Savins is helpful in that it shows the importance of the relative magnitudes of Wi and Re in assessing the pattern of the secondary flow.

The instability of the secondary flow pattern is brought about not only by the variation of the speed of the impeller but, even more so, by the variation of the liquid properties. Ide and White[20] have shown how the rheological properties of a styrene polymerizing solution change during bulk polymerization (Fig. 7). Between 20% and 50% conversion of monomer styrene, the zero-shear viscosity of the solution increases by a factor of three hundred and the zero-shear primary normal stress coefficient rises ten thousand times. As the initially low viscosity monomer styrene was converted into a highly viscous solution of polystyrene in styrene, the flow pattern in the reactor kept changing considerably. Several complete reversals took place during the polymerization and the interested reader is advised to read the original work cited above. In this context we only want to stress the difficulty of controlling the molecular weight distribution in a reactor with segregated zones which do not exchange mass.

Not much is known about the influence of viscoelasticity on the flow pattern around class (ii) impellers. While anchors and gates are notoriously known for producing a very poor axial circulation of the liquid content of the tank, Peters and Smith[14] found experimentally that the liquid's elasticity actually enhances the axial flow. Most of this axial flow takes place alongside the blades leading to an even number of toroidal circulations. In order to gain a better understanding of this phenomenon one would have to have a more

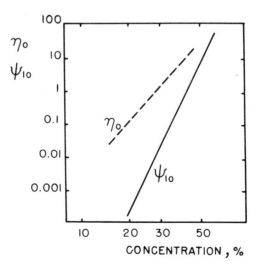

Figure 7   Zero-shear viscosity η_0 and primary normal stress coefficient ψ_{10} (both in S.I. units) of a polystyrene-styrene polymerizing solution (from Ide and White[20]).

accurate picture of the primary flow and of the associated deformation rates. This is, so far, lacking.

An opposite conclusion was arrived at independently by Chavan
et al.[21, 27] and by Carreau et al.[22] for class (iii) impellers, namely for
helical ribbons. The axial circulation is considerably reduced in
elastic liquids. Figure 8 shows typical data of axial velocities
obtained by Carreau et al.[22] for a helical ribbon. The axial velocities
in a 2% CMC solution (shear-thinning but of negligible elasticity)
are of the same magnitude as in Newtonian liquids. In a 1%
polyacrylamide solution of comparable apparent viscosity but
strongly elastic, the axial circulation is considerably damped: the
magnitude of the axial velocities is less than 2% of the blade tip
velocity. On the other hand, the tangential velocities were greatly
increased to the point that the whole content of the vessel (with the

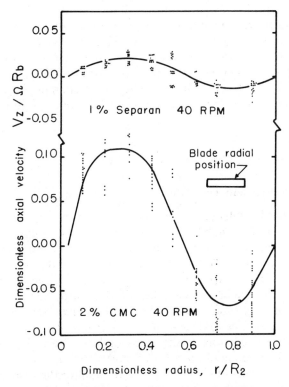

Figure 8   Axial velocity distribution in a vessel agitated by a helical ribbon impeller
(from Carreau et al.[22]).

exception of a thin layer near the walls) rotated as a solid body with an angular velocity equal to that of the impeller.

Because of the lack of any direct experimental evidence we may only speculate about the influence of the rheological complexities on the tertiary flow pattern, that is, in the stagnation region and in the vortices behind the blades. It is probably safe to say, that the shear-dependent viscosity is not likely to change qualitatively the picture of flow in and out of the vortices. Because of the high deformation rates in the vortex, however, the apparent viscosity will drop to a very low value and this will have some impact on processes controlled by shear deformation.

Recent experimental evidence, however, suggests that the maximum shear rate on the upstream face of a blade is higher in a shear-thinning liquid than in a Newtonian liquid. Wichterle et al.[12] measured the local shear rates by an electrochemical technique and found that the ratio $\dot{\gamma}/N$ increased sharply with the decrease of the flow index n.

Viscoelasticity, if manifested by an increase of the elongational viscosity, is likely to have a dramatic influence on the flow in the vortices. From the work of Fruman and Tulin[23] we are aware of the suppression of wakes and cavitations behind fins when a viscoelastic polymer is added to the bulk flow. The flow past an impeller blade is of the same nature and, therefore, similar results can be expected.

4.   MIXING AND CIRCULATION TIMES

4.1.   Definitions and Techniques

Although the knowledge of the overall flow pattern in a stirred tank is, in general, very helpful in understanding the operation of the mixer it is not easily convertible into a quantitative measure of mixing performance. To this end, the engineer would prefer a single parameter open to experimental measurement and uniquely linked to the performance of a mixer. The criterion most widely used is the so called "mixing time", t_M which will be defined here as the time required for a tracer to disperse or for a chemical reaction to reach an equilibrium. Obviously, the rates of both these processes are exponential and none of them is ever completed in a finite period of time. Also the scale of scrutiny chosen to monitor the response after the tracer was injected or the chemical reaction initiated will, to a large extent, affect the mixing time measurements. These two inherent drawbacks have often been cited to criticize the concept of mixing time. Although this criticism is well founded, mixing time continues to be widely used as a measure of mixing rate. This will be justified in all those cases where the mixing time is used only for comparison provided that the same experimental technique is used and the same end point on the response curve chosen.

When working with non-Newtonian liquids, we must be careful not to choose an experimental technique which would bring about changes in the rheological properties of the liquids mixed. This would be the case if pH or acid-base reaction was used to monitor the rate of mixing in aqueous solutions of polyelectrolytes. Ford et al.[24] reviewed this problem and suggested redox reactions, such as the reduction of iodine by ammoniumthiosulphate in the presence of modified starch, as being relatively safe. This conclusion has been confirmed later by a number of other researchers.

Whichever technique is used to monitor the mixing time, the response curve will show a distinct periodicity. This is brought about by consecutive passages of a cloud with a locally high concentration of the tracer or of the reactants. The time period between two subsequent peaks or troughs on the response curve is known as the "circulation time", t_C. Other techniques may be used to measure the circulation time, such as monitoring the passages of a freely suspended flow follower and averaging over a large number of observations.

It must be realized that the circulation time is not, strictly speaking, a measure of mixing since it reflects only the intensity of the macro-flow, that is, of the primary (tangential) and secondary (radial-axial) flows. This may be satisfactory if working with helical agitators but not so when radially thrusting impellers (such as turbines, flat blades, and anchors) are used. The tertiary flow (the stagnation flow on the up-stream face and the vortex flow on the down-stream face of the blade), where the shear and stretch rates are highest, does not contribute significantly to the variation of circulation times.

Mixing and circulation times are usually reported in a dimensionless form as Nt_M and Nt_C, respectively. Since the circulation time is related to the pumping capacity of the impeller q and to the volume of the liquid in the tank V, a circulation number Ci is sometimes defined as

$$Ci = q/Nd^3 \simeq V/t_C Nd^3 \qquad (19)$$

Axially thrusting impellers (helical ribbon, helical screw, and their combinations) are known to be much more efficient agitators for viscous liquids[25] than those pumping the liquid radially and this is true even more for viscoelastic liquids because of the complicated flow pattern (see preceding section for details). Therefore, most of the following discussion will deal with helical agitators.

4.2. Shear-Thinning Effects

It follows from the work of Chavan et al.[21, 26, 27] and Carreau et al.[22] that the shear dependence of the viscosity of non-Newtonian liquids has very little influence on the pumping capacity of helical impellers and, thus, on circulation times. Therefore, in the viscous regime of flow (Re < 10), the dimensionless circulation time is independent of the Reynolds number. In the intermediate regime of flow (10 < Re < 1000), however, Ford and Ulbrecht[28] observed that the dimensionless circulation time for a helical screw rotating in a centrally positioned draft tube decreased with increasing Reynolds number, but increased with the shear-thinning properties of the liquids.

A recent work of Guerin et al.[29] based on the observation of a freely suspended particle indicates, moreover, that the distribution of circulation times may be affected considerably. The experimental data for a helical-ribbon impeller are shown in Fig. 9 in which the

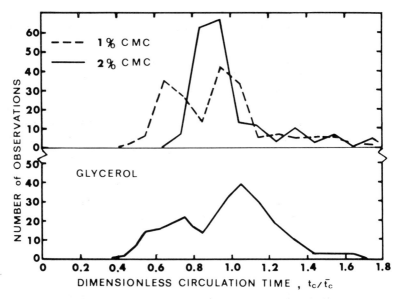

Figure 9   Distribution of circulation times for a helical ribbon impeller (from Guerin et al.[29]).

number of circulations having the duration t_c is plotted against the ratio $t_c/\bar{t}_C$, $\bar{t}_C$ being the average circulation time. In the case of glycerol and 1% aqueous CMC solution the distribution is broad and bimodal. The peaks observed at shorter circulation times correspond to short-circuited circulation loops, that is, to those cases when the particle moving upwards in the centre of the vessel swung to the vessel wall before reaching the liquid surface zone. In the case of the 2% CMC solution, the distribution is narrower with a single peak and this is characteristic of a strongly preferential circulation path. Because of the shear-thinning, the viscosity of the 2% CMC solution is much lower in the high shear region, that is, near the impeller blade and this leads to an effective segregation between the high circulation zone and the nearly stagnant zone. Although the average circulation time is as large as it is in a Newtonian liquid there is insufficient transfer of fluid between the high shear and low shear regions and the overall mixing efficiency is likely to be lower.

Similar conclusions can be drawn from the above cited reports about the influence of shear-thinning on mixing times, namely that the dimensionless mixing time, Nt_M, is constant in the viscous regime of flow. For helical screw impellers in the intermediate regime ($10 \leqslant Re < 1000$), Nt_M decreases with increasing Reynolds number, the shear-thinning effects being incorporated in a generalized Reynolds number[28]. For helical ribbon impellers, however, Carreau et al.[22] obtained mixing times considerably longer (2 to 3 times) in shear-thinning CMC solutions than in Newtonian liquids although the circulation times were comparable. Dimensionless mixing times for five helical ribbons of different geometries are reported in Table 1. It can be observed that the mixing rate in a shear-thinning inelastic CMC solution depends strongly on the geometrical configuration of the impeller. The very long mixing times observed with the smaller diameter impeller IV (which has a larger gap between the blades and the vessel wall) is attributed to the presence

Table 1 Dimensionless Mixing Time of Helical Ribbon Agitators[a]

Impeller	n_b	D/d	P/d	W/D	Nt_M Glycerol	2% CMC	1% Separan
I	2	1.11	0.72	0.10	45	105	125
II	2	1.11	1.05	0.10	51	107	142
III	2	1.11	0.71	0.20	25	51	108
IV	2	1.37	0.85	0.12	55	189	137
V	1	1.11	0.70	0.10	61	120	163

[a] In all cases, $D/H = 1$. For details of the geometries, see Patterson et al.[46].

Figure 10   Conical stagnant zone observed in a 2% aqueous solution of sodium carboxy methyl cellulose with a helical ribbon agitator rotating in clockwise direction. The photograph was taken during the decoloration process used to measure mixing time (from Yap[47]).

of nearly stagnant zones as discussed above. A typical stagnant zone is illustrated in Fig. 10 for a 2% CMC solution. Such a stagnant zone observed in the dark cone at the bottom of the vessel could persist for hours when rotating the impeller in the clockwise direction (pumping upward at the blades). In the counter-clockwise direction, the stagnant zones are not as important but nevertheless they affect the mixing rate in shear-thinning liquids. This drastic influence of the flow direction is not yet clearly understood.

4.3.   Elastic Effects

The influence of viscoelasticity on both circulation and mixing times is much more dramatic. Chavan et al.[21] measured separately the angular and the axial circulation times in a viscoelastic liquid stirred by a combined helical ribbon-helical screw agitator and found that, while the axial circulation was reduced to almost zero, the angular circulation was enhanced to the point where most of the liquid content of the tank rotated with the angular velocity of the impeller as a solid body. Carreau et al.[22] found the same for a helical ribbon. They found the angular velocities in a large section of the tank almost equal to the velocity of the blade while the axial velocities were close to zero (see Fig. 8). These findings are particularly striking since the helical agitators are generally considered best for generating intensive axial circulation.

On the other hand, the presence of elasticity improved the axial circulation in a vessel stirred by an anchor agitator[14] which is normally expected to produce mainly angular circulation with very little axial flow. These two findings illustrate well how the rheological complexities are coupled with the geometry of flow and how conclusions arrived at by extrapolations from one geometry to another could be misleading.

If the circulation times in vessels stirred by helical agitators are increased by the presence of viscoelasticity, then the mixing times are also increased but not necessarily by the same factor. In an attempt to quantify the influence of viscoelasticity, Ford and Ulbrecht[28] correlated the circulation and the mixing times in a tank stirred by a helical screw impeller operating in a draft tube with the Weissenberg number. They obtained

$$(Nt_C)_{nN}/(Nt_C)_N = (1 + 0.45 \, \text{Wi})^{0.3}(\eta_0/\eta_e)^{0.3} \tag{20}$$

$$(Nt_M)_{nN}/(Nt_M)_N = (1 + 0.45 \, \text{Wi})^{0.8} \tag{21}$$

where the subscripts nN and N refer to non-Newtonian and Newtonian fluids respectively. The Weissenberg number is defined here as

$$\text{Wi} = (\psi_{10}/\eta_0)N \tag{22}$$

where ψ_{10}/η_0 is the zero-shear limit of ψ_1/η. The ratio of the zero-shear viscosity to the effective viscosity in Eq. (20) accounts for shear-thinning effects on the circulation times. The dimensionless mixing time is not influenced by shear-thinning but increases faster with increasing Weissenberg number than the circulation time. Most viscoelastic liquids, however, are also shear-thinning and the overall effect may, in general, be similar.

For helical ribbon impellers, Carreau et al.[22] found that the dimensionless mixing time increased by a factor of 2 to 7 with increasing shear-thinning and elastic properties. Although the viscoelastic liquids were more difficult to mix, it was not possible to distinguish between elastic and shear-thinning effects. Moreover, the authors observed that the influence of rheological properties was considerably different from one geometry to another one. Table 1 compares the dimensionless mixing times obtained with five helical ribbons of different geometries in three liquids of different rheological properties. Except for geometry IV (wider gap), the mixing times are considerably longer in the elastic polyacrylamide solution (1% Separan). In the case of impeller IV, the liquid's elasticity improves its mixing performance in non-Newtonian liquids. As reported above, the vertical circulation in elastic liquids is reduced to almost zero, hence mixing with helical ribbon agitators proceeds mainly through shear deformation and large local velocity fluctuations.

4.4. Blending of Non-Newtonian Liquids

Blending of liquids with different rheological properties is a common situation in industrial operations. Ford and Ulbrecht[30] measured the blend times of two equal volumes of aqueous polymer solutions using a helical screw in draft-tube mixer. Both inelastic as well as viscoelastic solutions were used. The variation of the zero-shear viscosity and of the primary normal stress coefficient exceeded five decades. Apart from varying the rheological properties also the orientation of the two liquids was varied with the more viscous liquid being either downstream or upstream from the less viscous liquid in the draft tube. Although these solutions were, thermodynamically, completely miscible the blend times were up to five times longer than in liquids of uniform properties. Also the viscoelasticity increased the blend time. It was found empirically that the dimensionless blend time is proportional to

$$Nt_B \sim (1 + 0.52 \, \text{Wi})^{1.25} \qquad (23)$$

where the Weissenberg number is defined by Eq. (22). As to the influence of the orientation it was found that it is more difficult to disperse a less viscous fluid into a more viscous fluid than the other way round. The dimensionless blend time is proportional to

$$Nt_B \sim (\eta_{0M}/\eta_{0L})^z \qquad (24)$$

where $z = 0.059$ when the more viscous fluid (η_{0M}) is initially in the upstream location in the draft-tube and $z = 0.17$ when the less viscous fluid (η_{0L}) takes this position. The full correlation is given in a graphical form by Ford and Ulbrecht[30] and by Skelland[6].

4.5. Non-Newtonian Effects for Class (i) Agitators

Much less is known about the mixing and circulation times in vessels stirred by turbines, paddles and propellers. An early work by Norwood and Metzner[31] suggests that correlations derived for Newtonian liquids could be employed provided that a generalized Reynolds number is used. However, the mixing time data of Godleski and Smith[32] are ten to fifteen times larger than those predicted by the Norwood–Metzner correlation and this seems to confirm that a degree of segregation exists between the high shear rate (low apparent viscosity) and the low shear rate (high apparent viscosity) zones. In highly shear-thinning liquids (showing an effective yield stress), a cavern of turbulent flow will surround the fast moving impeller while the rest of the liquid batch might be at rest[33]. A formula to predict the size of such a cavern was proposed by Solomon et al.[34] It is obvious that, under these circumstances, the concept of mixing and circulation times has no meaning because no mixing (except for molecular diffusion) takes place between the cavern and the rest of the vessel.

A similar deterioration of mixing performance may be expected with viscoelastic liquids in view of what has been said in the preceding section about the flow reversal and the formation of segregated zones. The photographs of Giesekus[35] clearly show that a tracer injected in the inner region surrounding the turbine impeller will stay there almost indefinitely being transported across the zero stream plane only by molecular diffusion. These results, however, must not be applied to mixers of different geometry. It has been already pointed out that Peters and Smith[14] found an increase of axial circulation due to viscoelasticity in vessels stirred by an anchor agitator. They also found that the mixing and the circulation times were reduced in this case.

5. POWER CONSUMPTION

The interest in power consumption during mixing stems not only from economic considerations but also from the fact that changes in the torque are indicative of changes in the overall flow pattern in the tank. Also power measurements are frequently used to characterize the liquid's rheology in a mixing system.

In this section, we shall deal first with the influence of the shear thinning and then, separately, with the influence of viscoelasticity on the power needed to maintain an agitator at a constant speed.

The dimensionless analysis[6] will show that, in the absence of static head variations, the power number Po

$$Po = P/d^5N^3\rho \qquad (25)$$

will be a function of the Reynolds number

$$Re = d^2N\rho/\mu \qquad (26)$$

in mixers which are geometrically similar. The available experimental evidence shows that the product of the power number and of the Reynolds number depends only on the geometry of the mixer in the viscous regime of flow (Re < 10) for Newtonian liquids.

5.1. Shear-Thinning Effects

With shear-thinning liquids, however, we face the difficulty of choosing the correct effective viscosity η_e for the Reynolds number since, as the shear rate varies across the gap between the impeller and the vessel wall, the viscosity increases from its minimum value in the close vicinity of the impeller to its maximum value in regions far away from the impeller. Following the idea of Magnusson[36], Metzner and Otto[2] suggested that the effective viscosity should be evaluated at an effective shear rate $\dot{\gamma}_e$ which is proportional to the rotational speed of the impeller.

$$\dot{\gamma}_e = k_s N \qquad (27)$$

This assumption was subsequently verified by the measurements of local velocity gradients reported by Metzner and Taylor[3]. They showed that the local shear rate in solutions of carboxy methylcellulose (CMC) varied linearly with the speed of the impeller and decreased rapidly with increasing distance from the impeller. Metzner's approach is now the accepted technique for the design of impellers of class (i) in the viscous regime (Re < 10) of mixing with the coefficient k_s being around 11. An extensive tabulation of k_s values as well as a detailed algorithm for the computation of the power number is given by Skelland[6].

Once the effective shear rate has been established, the corresponding effective viscosity can be found either from experimental viscosity data or by using an appropriate rheological model. In the past, the power-law expression has been used most frequently with satisfactory results. Recently, Ducla et al.[37] confirmed the usefulness of this design technique for low shear rates for which the Cross–Williamson model, Eq. (11), is more appropriate than the "Power-Law".

There are, however, some doubts as to whether this method is applicable to very strongly shear-thinning liquids, that is, to those which may be showing an effective yield stress. It has been already pointed out in the preceding section that, in such a case, only parts of the liquid batch may be flowing while the rest of the tank content is stationary.

At this point it needs to be emphasized that the effective shear rate defined by Eq. (27) relates only to the power consumption in the stirred tank and it should not be used to calculate heat transfer to coils or vessel walls or to estimate drop or bubble size. Van't Riet and Smith[38] found, using their photographic data, that the shear rate in the vortices behind the blades is up to ten times higher than that determined by Eq. (27). More recently, Wichterle et al.[12] reported shear rates on the upstream face of an impeller blade measured by a polarographic method. They, too, found the shear rate about five times higher in the viscous regime of flow (Re < 10) and ten times higher in the transition regime (Re > 10). Further, they showed that the maximum shear rate increases rapidly with the increase of the shear thinning (i.e., with the decrease of the flow index) so that the maximum shear rate could be two orders of magnitude above that predicted by Eq. (27) for highly shear-thinning systems, such as suspensions of kaolin. These results, however, are not necessarily in contradiction since the effective shear rate of Eq. (27) represents an average value over the whole tank.

For the class (ii) agitators, such as anchors and other close clearance impellers, the coefficient k_s in Eq. (27) is no longer a constant but it becomes function of both the width of the gap between the blade and the vessel wall and the rheological properties of the liquid. Calderbank and Moo-Young[39] proposed for shear-thinning liquids

$$k_s = \left[9.5 + \frac{9(D/d)^2}{(D/d)^2 - 1} \right] \left(\frac{4n}{3n + 1} \right)^{n/(1-n)} \tag{28}$$

A close examination of this equation will reveal that the dependence of k_s on the flow index n is rather weak[40].

Turning now to class (iii) impellers, an alternative approach was developed by Chavan and Ulbrecht[26, 41-43, 45] based on the idea of Bourne and Buttler[44] that the flow between the rotating helical surface and the vessel wall can be simulated by a Couette flow. Chavan and Ulbrecht improved the concept by defining an equivalent cylinder experiencing the same overall friction as the helical surface of the ribbon and of the screw respectively and by extending this model to the "power-law" liquids. It follows then from their analysis for the helical screw impeller rotating in a draft tube (see Fig. 11 for nomenclature) that

$$Po = \frac{\pi A}{2} \left(\frac{d_e}{d}\right)\left(\frac{D_t}{d_e}\right)^2 \left\{\frac{4\pi}{n[(D_t/d_e)^{2/n} - 1]}\right\}^n \left(1 + \frac{d}{c}\right)^{0.37}$$

$$\times \left(\frac{D - D_t}{h_t}\right)^{-0.046} \left(\frac{C_r}{d}\right)^{-0.036} \left(\frac{d^2 N^{2-n} \rho}{m}\right)^{-1} \tag{29}$$

where the equivalent diameter is given by

$$\frac{d_e}{d} = \frac{D_t}{d} - \frac{2W}{d} \bigg/ \ln\left\{\frac{(D_t/d) - 1 + 2(W/d)}{(D_t/d) - 1}\right\} \tag{30}$$

and the dimensionless surface is

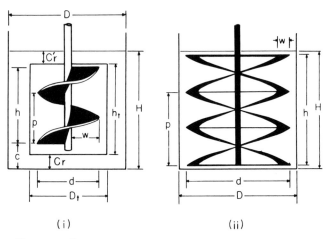

(i)   (ii)

Figure 11   (i) Helical screw impeller. (ii) Helical ribbon impeller.

$$A = \frac{(p/d)(h/d)}{3\pi} \left[\frac{\pi \sqrt{(p/d)^2 + \pi^2}}{(p/d)^2} + \ln \left(\frac{\pi}{p/d} + \frac{\sqrt{(p/d)^2 + \pi^2}}{(p/d)} \right) \right]$$
$$\times \{1 - [1 - 2(w/d)]^2\} + \pi[1 - 2(w/d)](h/d) \tag{31}$$

For helical ribbon impellers, Chavan and Ulbrecht[26, 42] found

$$\mathrm{Po} = 2.5 \, A\pi \left(\frac{d_e}{d} \right) \left(\frac{D}{d_e} \right)^2 \left\{ \frac{4\pi}{n[(D/d_e)^{2/n} - 1]} \right\}^n \left(\frac{d^2 N^{2-n} \rho}{m} \right)^{-1} \tag{32}$$

where

$$\frac{d_e}{d} = \frac{D}{d} - 2\frac{W}{d} \Big/ \ln \left\{ \frac{(D/d) - 1 + 2(W/d)}{(D/d) - 1} \right\} \tag{33}$$

and

$$A = \frac{(h/d)(p/d)}{3\pi} \left[\frac{\pi \sqrt{(p/d)^2 + \pi^2}}{(p/d)^2} + \ln \left(\frac{\pi}{p/d} + \frac{\sqrt{(p/d)^2 + \pi^2}}{p/d} \right) \right]$$
$$\times \{1 - [1 - 2(W/d)]^2\} \tag{34}$$

The power is doubled if the impeller has two identical ribbons.

The above expression holds also for combined helical ribbon-screw agitators provided that the area of the screw is smaller or equal to the area of the ribbon. In a parallel publication, Chavan and Ulbrecht[45] developed the model also for an off-centre helical screw impeller operating without a draft tube.

Equations (29) and (32), which have no adjustable parameters, were tested using experimental data published by twelve research groups covering 50 different geometrical configurations and several shear-thinning liquids (some of them were also viscoelastic) in the viscous regime of flow ($\mathrm{Re} < 10$). The average deviation for the helical screw in a draft tube was found to be 7% and for the helical ribbon 15%.

A different model based on the drag flow past a helical ribbon was developed for Newtonian liquids by Patterson et al.[46] in which the analogy with the Couette flow was not required. The following correlation was obtained

$$\mathrm{Po} = 24 n_b [\mathrm{Re}^{0.93} (D/d)^{0.91} (d/l)^{1.23}]^{-1} \tag{35}$$

where n_b is the number of blades and l is the total length of a blade.

It is interesting to note that the exponent of the Reynolds number in Eq. (35) is slightly larger than -1, a value typical for the viscous regime of flow. The data of Patterson et al. were obtained in the range of Reynolds numbers between 10 and 800, in which one would expect to note an influence of the inertial forces. The correlation (35) was extended by Yap et al.[47] to include shear-thinning and viscoelastic liquids, using the concept of effective viscosity and taking

$$k_s = (4)^{1/(1-n)}(d/D)^2(l/d) \tag{36}$$

Like in the case of Eq. (28), the coefficient depends on the geometrical configuration of the impeller but, unlike for class (ii), it is a strong function of the flow index n, varying from 20 to almost 200. This is in line with an earlier work of Prokopec and Ulbrecht[48] who applied the effective viscosity approach to a helical screw agitator in a draft tube and found an average value of k_s to be close to one hundred and strongly dependent on the flow index of the liquid tested. Clearly, the very concept of the effective viscosity is hardly useful in this case.

In an attempt to clarify the influence of the shear-thinning properties on the effective shear rate in a mixing vessel, Carreau[49] proposed a simple analysis based also on the analogy with the Couette flow. As for the models of Chavan and Ulbrecht[26, 41-43, 45], the impeller is simulated by an inside coaxial cylinder of equivalent diameter d_e but the rheological properties and the torque are evaluated at the vertical wall of the vessel (the contribution to the torque of the bottom wall is assumed negligible). For power-law liquids in the viscous regime of flow, the effective shear rate is proportional to the rotational speed of the impeller but the proportionality constant depends on the geometrical configuration of the impeller and on the power-law index n:

$$\dot{\gamma}_e/N = k_s = \left[\frac{k_p d^3}{\pi^2 D^2 H}\right]^{1/(1-n)} \left[\frac{n[(D/d_e)^{2/n} - 1]}{4\pi}\right]^{n/(1-n)} \tag{37}$$

where the coefficient k_p is defined by the classical relation

$$Po = k_p \, Re^{-1} \tag{38}$$

and can be determined by power measurements in Newtonian liquids. The equivalent diameter d_e is determined with the help of the relation obtained for Newtonian liquids

$$k_p = \frac{4\pi^3 D^2 H}{d^3[(D/d_e)^2 - 1]} \tag{39}$$

We note that Eq. (37) can be derived also from the models of Chavan and Ulbrecht[26, 41-43, 45] presented above. Although the value of k_s seems to be strongly dependent on the power-law index, it is not the case in the two extreme examples discussed by Carreau[49]. For a turbine impeller (with k_p equal to 70) the value of k_s increases from 6 to 11 as the power-law index decreases from 0.9 to 0.1. This is more or less in line with values reported in the literature[6, 40]. On the other hand, using a value of k_p equal to 400 for a helical ribbon impeller, no significant variation of k_s with the power-law index was observed and the average value of 35 is quite close to the value of 30 reported by Nagata[50] for a slightly different helical ribbon agitator.

Equation (37) is not capable of describing large shear-thinning effects on the effective shear rate as observed experimentally by Yap et al.[47] for helical ribbon agitators and expressed by equation (36). However, the very large values of k_s were obtained for mildly shear-thinning liquids (CMC solutions) which obey the power-law expression under a very limited range of shear rate. Moreover the power number (or the effective viscosity) is not sensitive to the value of k_s as n approaches the value of one. Hence small errors in the power or the viscosity measurement with mildly shear-thinning liquids could lead to large errors in the determination of k_s.

5.2. Influence of Flow Regime

Bourne et al.[51] have made a simple analysis to show that the effective shear rate is no longer proportional to the impeller rotational speed when mixing does not proceed in the viscous regime of flow. Their result is, however, incomplete since they make use of the usual generalized Reynolds number that incorporates a constant coefficient k_s. This is in contradiction with their findings. Carreau[49] extended the analysis presented above to the case where the power number is no longer proportional to the inverse of the Reynolds number. Under the assumption that the effective shear rate in the transition regime is related to the shear rate at the vessel wall by the same expression as in the viscous regime (this assumption appears reasonable for mixing regimes not too far from the viscous one), the effective shear rate is given by

$$\dot{\gamma}_e = \left[\frac{k_p d^3}{\pi^2 D^2 H}\right]^{1/r(1-n)} \left[\frac{n[(D/d_e)^{2/n} - 1]}{4\pi}\right]^{n/r(1-n)} \left[\frac{d^2\rho}{m}\right]^{(1-a)/r} N^{(2-a)/r}$$

(40)

where a is the exponent affecting the Reynolds number in the power correlation

$$Po = k_p/Re^a = kp \left[\frac{m\dot{\gamma}_e^{n-1}}{d^2 N\rho}\right]^a$$

(41)

and $r = n(1 - a) + a$. Obviously for $a = 1$, this result reduces to Eq. (37). For $a < 1$, it is different from the result of Bourne et al.[51] but it predicts similar effects: the effective shear rate is no longer proportional to the rotational speed of the impeller and it depends strongly on the scale of the impeller and on the liquid's properties.

The data of Bourne et al.[51, 52] for a 63% (per weight) suspension of calcium carbonate in ethylene glycol stirred by anchor agitators were used to test the validity of Eq. (40). For $a = 0.6$ and $n = 0.64$, the effective shear rate is found to be proportional to

$$\dot{\gamma}_e \sim d^{0.93} N^{1.63}.$$

The data and the predictions are compared in Fig. 12. The data for

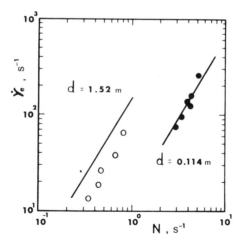

Figure 12   Effective shear rate in the transition regime of flow for a 63% suspension of calcium carbonate in ethylene glycol mixed by anchor impellers. ○● Data of Bourne et al.[51] —Predictions of Eq. (40).

the small-scale system have been used to calculate the proportionality constant. This result describes well the dependence of the effective shear rate on the impeller rotational speed but the scale effect is, however, somewhat overestimated. Although this analysis needs to be verified with more data, it shows clearly that the Metzner-Otto concept is no longer valid in the transition regime of flow. The use of a constant k_s could lead possibly to very large errors when designing large-scale equipment. Considering the many industrial mixing operations with non-Newtonian liquids proceeding in the transition regime (mainly for class (i) agitators), this area of research is worth further investigation.

In the turbulent regime of flow (Re > 1000) the power number does not depend on the Reynolds number and the concept of the effective viscosity does not apply. Here, as well as in the transition regime, the safest design method is that of using empirically obtained plots of power number[6].

5.3. Elastic Effects

The influence of the liquid's elasticity on power consumption in the viscous regime of flow is far from clarified but, in any event, it is not likely to be large since the primary flow pattern which is mainly responsible for the energy dissipation is hardly altered by the presence of elastic forces. Chavan and Ulbrecht[26, 41-43, 45] and Yap et al.[47] correlated successfully the power data for shear-thinning visco-elastic polymer solutions stirred by helical agitators (both screws and ribbons) without accounting specifically for the liquid's elasticity.

Nienow et al.[13] reported slightly increased power consumption when stirring viscoelastic xanthan gum solutions with a standard turbine impeller that, they concluded, would lead to an increase of the Metzner–Otto coefficient should all data be brought on the same correlation line. On the other hand, Ducla et al.[37] derived a somewhat lower value of the same coefficient from the power measurement of turbine impellers stirring viscoelastic polymer solutions. They did not, however, report the power data nor did they measure the primary normal stress coefficient (the liquid's elasticity was evaluated from stress relaxation experiments). There is, however, no obvious conflict between these observations because the existence of viscoelasticity in itself does not result unambiguously into either a power increase or decrease. The power consumption is given only by the rate of energy dissipation and this, in turn, depends on the overall flow pattern. Since the primary flow pattern is not altered substantially by the viscoelasticity we have to look for the explanation in the mode of the secondary and tertiary flow patterns. With reference to section 3 of this chapter, at some values of Wi/Re, the viscoelasticity reduces the extent of the secondary circulation and this, in turn, results in a lower power consumption. At other values of this criterion, the elasticity leads to the creation of dual, mutually opposing circulations which dissipate more energy so that the power consumption will be larger.

Further, it needs to be realized that most polymer solutions are not only viscoelastic but also shear-thinning so that the observations reported above are influenced by both rheological complexities.

In order to separate the role of elasticity from that of shear dependent viscosity, Kelkar et al.[17] measured the torque of turbines rotating in viscous silicon oils (polysiloxanes) which are viscoelastic but the viscosity of which is independent of shear rate (from 45 to 65 Pa.s). They found that the product $C_M \cdot Re$ is almost constant for We < 1 but it decreases as the Weissenberg number increases from 1 to 20 (see Fig. 13). This decrease of the power consumption coincides with the suppression of the inertia driven secondary flow due to the increase of the elastic forces. It is plausible to conjecture that the curve would reach a minimum and then start rising again as the secondary flow is reversed. Eventually, one would expect the power number to be higher than that for an inelastic liquid as the two segretated regions of secondary flows are formed and more energy is needed to drive them. This conjecture is substantiated by a minimum on a mass transfer curve observed in viscoelastic liquids by Deslouis and Tribollet[53]. Recently, Prud'homme and Shaqfeh[54]

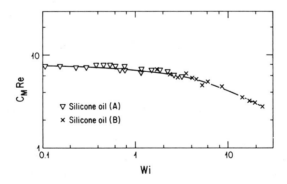

Figure 13   Power consumption in viscoelastic silicone oils for a standard Rushton turbine impeller (from Kelkar et al.[17]).

repeated Kelkar's experiments using a so called Boger liquid (small amounts of high molecular weight polyacrylamid dissolved in corn syrup) and found a much higher power number than that in an inelastic corn syrup. However, no parallel observation of the flow pattern has been given. Obviously more work is needed to explore this relatively unmapped area.

When the concentration of polymer additives (or products) is low (say less than 1000 ppm) the viscosity of such solutions is close to that of the solvent (and almost shear independent) but the elastic phenomena might still be quite significant in the transition or turbulent regime of flow. All the available experimental evidence[13, 55, 56], confirms substantial reductions of torque. This phenomenon can be best interpreted by considering the energy dissipation in the tertiary (vortex) flow behind the impeller blades which, along with the macrocirculation, contributes to the power consumption. The presence of these so called "drag reducing" polymers limits the lower end of the turbulent spectrum leading thus to the reduction of the turbulent energy dissipation. This phenomenon should not be interpreted as being unequivocally advantageous for industrial applications since the reduction of momentum transport is accompanied by a reduction of convective[57] and interfacial[58] mass transport. A similar reduction of heat transfer is also expected.

6. MIXING EFFICIENCY

A useful criterion for comparing the performance of various impellers in liquids of different rheological properties is the total energy per unit volume of liquid required to achieve a desired degree of homogeneity. Such a criterion was used by Patterson et al.[46], Yap et al.[47], and Guerin et al.[29] for comparing the performance of helical agitators. In the viscous regime of flow, the criterion can be defined as

$$(\text{Eff})_{\text{rel}} = 1000\eta_e ND^2 H/P \cdot t_M \qquad (42)$$

The number 1000 is arbitrary and the relative efficiency is constant for a given geometry in Newtonian liquids. In non-Newtonian liquids, the decrease of the efficiency is uniquely related to the increase of the mixing time t_M as the use of the effective viscosity η_e accounts for possible rheological effects on the power consumption.

The changes of both the mixing time and the power consumption with the geometrical configuration of the mixer are included in the relative efficiency. Patterson et al.[46] used d^3 instead of D^2H, thus they introduced an undesirable dependence on the impeller diameter when comparing mixers of different geometrical configurations.

Guerin et al.[29] compared with this criterion the efficiency of a screw impeller rotating in a draft coil and of five helical ribbons in Newtonian liquids. They showed that the screw impeller performs better than the best helical ribbon. On the other hand, Yap et al.[47] compared the efficiency of the same helical ribbons when mixing non-Newtonian liquids. The relative efficiencies have been recalculated using Eq. (42) and they are reported in Table 2. The efficiency is approximately two times lower in a shear-thinning CMC solution and more than four times lower with the wider-blade impeller III in a highly elastic polyacrylamide (Separan) solution although this impeller was found to be quite efficient in Newtonian and shear-thinning inelastic liquids. On the other hand, because of the much lower torque the smaller diameter impeller IV (wider gap between the blades and the vessel wall) performs considerably better than the others in Newtonian and in viscoelastic liquids but its performance in the CMC solution may not be acceptable with very long mixing times (see Table 1). As mentioned previously, the presence of nearly stagnant zones when mixing CMC solutions is responsible for the much longer mixing times with this geometry. Yap et al.[47] have reached a somewhat different conclusion because they used an efficiency criterion which is dependent on the impeller diameter. These results stress the necessity of using different rules when designing mixing equipments for liquids of different rheological properties. As shown by Table 2, the performance in a given

Table 2   Relative Efficiency of Helical Ribbon Agitators[a]

Impeller	n_b	D/d	P/d	W/d	(Eff.)$_{rel}$ Glycerol	2% CMC	1% Separan
I	2	1.11	0.72	0.10	0.075	0.040	0.030
II	2	1.11	1.05	0.10	0.137	0.070	0.052
III	2	1.11	1.05	0.20	0.188	0.081	0.041
IV	2	1.37	0.83	0.12	0.221	0.069	0.090
V	1	1.11	0.70	0.10	0.129	0.063	0.049

[a] Based on data of Yap[47]
    In all cases, $D/H = 1$. For details of the geometries, see Patterson et al.[46]

liquid can vary considerably and the best performing impeller in Newtonian liquids is not necessarily the optimum impeller for mixing shear-thinning and viscoelastic liquids.

Similar criteria have been used by Bourne et al.[59] and by Hoogendoorn and Den Hartog[60] to compare the efficiency of various impellers in Newtonian liquids. Another interesting comparison has been reported by Johnson[61]: at constant mixing time (300 s), the total energy to mix a corn syrup ($\mu = 2.4$ Pa.s) with a helical ribbon agitator was found to be six to twelve times less than with turbine and propeller agitators. It would be of interest to extend such a comparison to viscoelastic liquids. In view of the results reported in the preceding paragraph, the relative gain of helical impellers may be considerably less. This is another research area worth exploring.

If the work of Johnson shows that helical ribbons are efficient agitators for viscous liquids, Murakami et al.[62] have obtained a much better performance with reciprocating (up and down motion) impellers. At constant mixing time, the energy required to mix a Newtonian liquid ($\mu = 4.0$ Pa.s) with a reciprocating paddle was found to be ten times less than with a helical ribbon agitator. Similar efficiency increases are expected in non-Newtonian liquids as the up and down motion of the impeller should generate rapidly changing secondary flows favoring the interchange of liquid from the bottom and top parts of the vessel. Comparable results could possibly be achieved with simpler designs, as for example off-centre helical impellers.

7. CONCLUSIONS

This chapter has been confined to the mixing of non-Newtonian liquids in mechanically agitated vessels. Two important rheological anomalies, the shear-dependent viscosity and the primary normal stress differences observed in a simpler Couette flow have been used to illustrate how the liquid's rheological properties may affect the hydrodynamics in a mixing vessel.

The primary flow pattern created by an impeller is not considerably changed by the liquid's rheology whereas the secondary (vertical circulation) and the tertiary flow (such as the vortex flow behind the blades of the impeller) are markedly modified by the presence of elastic forces. Depending, however, on the class of agitators used the viscoelasticity may play a totally different role. The vertical flow circulation is drastically reduced when mixing viscoelastic liquids with helical impellers (screw and helical ribbon) whereas the same flow is enhanced by the use of an anchor impeller. Complete reversals of the secondary flow pattern are expected in vessels agitated by turbine impellers as the liquid becomes more and more elastic.

The mixing rate of helical screw impellers rotating in a draft tube is not affected by the shear-thinning properties of the liquid but decreases with increasing elasticity. Similar effects of elasticity are observed for helical ribbons but their mixing efficiency is also considerably decreased in shear-thinning inelastic liquids because of · the formation of nearly stagnant zones. Very long mixing times are obtained when mixing highly shear-thinning liquids with turbines, paddles and propellers, mainly in the viscous regime of flow, because of the segregation between the high shear rate region near the impeller and the very low shear rate zone in the rest of the liquid batch. A deterioration of the mixing performance is expected in viscoelastic liquids although it has been observed that the liquid's elasticity increases the mixing rate of anchor agitators.

The power consumption of an impeller mixing non-Newtonian liquids in the viscous regime of flow (Re < 10) can be estimated from the Newtonian power correlation with the help of the Metzner–Otto concept of effective shear rate. Specific models are also available to predict successfully the power consumption of helical agitators. In the viscous regime, the liquid's elasticity appears to have only a slight influence on power consumption. Mixing in the transition regime of flow needs to be fully investigated. In the light of little data and of a simple analysis, the effective shear rate is no longer proportional to the rotational speed of the impeller and it depends strongly on the liquid's properties and on the scale and geometrical configuration of the impeller. One may also expect that viscoelasticity affects considerably the power consumption in the transition as well as in the turbulent regime.

The total energy required per unit volume of liquid to achieve a desired degree of homogeneity can be used as a criterion to compare the efficiencies of different agitators. On that basis, the helical impellers have been shown to be much more efficient than turbines and propellers for mixing viscous Newtonian liquids. The efficiency of helical ribbons, however, is considerably reduced in non-Newtonian liquids and the best performing impeller in Newtonian liquids may not be the optimal impeller for mixing shear-thinning and/or viscoelastic liquids. New designs, such as reciprocating (up and down motion) impellers or off-centre helical impellers may prove to be much more efficient for handling rheologically complex liquids.

8. NOTATION

A dimensionless surface of helical impeller, Eqs. (31) or (34)

a parameter in the power number correlation, Eq. (41)

Ci   circulation number defined by Eq. (19)

c_r clearance between the screw impeller and the bottom wall, m

c_r' clearance between the draft tube and the bottom wall, m

c distance between the liquid's surface and the draft tube, m

c_M $4Po/\pi^4$, torque coefficient

D reservoir diameter, m

D_t diameter of draft tube, m

De   Deborah number defined by Eq. (18)

d impeller diameter, m

d_e equivalent diameter of impeller defined by Eqs. (30), (33) or (38), m

G elastic modulus, Pa

H height of liquid in the reservoir, m

h height of helical impeller, m

h_t height of draft tube, m

k_ρ proportionality constant of the power number

k_s Metzner–Otto coefficient defined by Eq. (27)

l total length of helical ribbon blade, m

m power-law parameter in Eq. (6), Pa.s^n

M torque exerted on cylinder, N.m

m' parameter in Eq. (14), Pa.s$^{n'}$

n power-law index in Eq. (6)

n' parameter in Eq. (14)

n_b number of helical ribbon blades

N impeller rotational speed, rev/s

N_1 primary normal stress differences, Pa

N_2 secondary normal stress differences, Pa

P power, W or pressure, Pa

Po   power number defined by Eq. (25)

p pitch of helical impeller, m

q vertical circulation flow rate, m^3/s

R_b radius of helical ribbon impeller, m

Re   Reynolds number defined by Eq. (26)

R_1 radius of inside cylinder, m

R_2 radius of outside cylinder or vessel, m

r parameter in Eq. (40)

t_B blending time, s

$\underline{t_C}$ circulation time, s

$\bar{t}_C$ mean circulation time, s

t_M mixing time, s
V volume of liquid, m^3
w blade width, m
Wi Weissenberg number defined by Eq. (17)

Greek Symbols

α parameter in the Ellis model, Eq. (10)
$\dot{\gamma}$ shear rate, s
$\dot{\gamma}_e$ effective shear rate, s
η non-Newtonian viscosity, Pa.s
η_e effective viscosity in a mixing vessel, Pa.s
η_0 zero-shear viscosity, Pa.s
η_∞ high-shear limiting viscosity, Pa.s
λ characteristic time of liquid's elasticity, s
λ_c characteristic time in Eqs. (9) or (11), s
μ viscosity of Newtonian liquids, Pa.s
μ_p viscosity in the Bingham model [Eq. (8)], Pa.s
ρ liquid's density, kg/m^3
τ_{21} shear stress, Pa

$\left.\begin{array}{l}\tau_{11}\\\tau_{22}\\\tau_{33}\end{array}\right\}$ normal stress components, Pa

τ_y yield stress parameter in Eqs. (7) or (8), Pa
$\tau_{1/2}$ parameter in the Ellis model [Eq. (10)], Pa
ψ_1 primary normal stress coefficient, defined by Eq. (12), Pa.s^2
ψ_2 secondary normal stress coefficient, defined by Eq. (13), Pa.s^2
Ω angular velocity of cylinder or impeller, s^{-1}
ω angular velocity of liquid, s^{-1}

REFERENCES

1. Moo-Young, M. B. (1969). The rheological effects of substrate additives on fermentation yields, *Biotechnol. Bioeng.*, **11**, 725; see also USP Nos. 3975236 and 3947323.
2. Metzner, A. B., and Otto, R. E. (1957). Agitation of non-Newtonian fluids, *AIChE J.*, **3**, 3.
3. Metzner, A. B., and Taylor, J. S. (1960). Flow patterns in agitated vessels, *AIChE J.*, **6**, 109.
4. Ulbrecht, J. J. (1974). Mixing of viscoelastic liquids by mechanical agitation, *The Chem. Eng. (London)* **286**, 347.
5. Ulbrecht, J. J. (1975). Influence of rheological properties on mixing, (in English), *Verfahrenstechnik*, **9**, 457.

6. Skelland, A. H. P. (1983). Mixing and agitation of non-Newtonian fluids, in *Handbook of Fluids in Motion* (Eds. N. P. Cheremisinoff and R. Gupta), Ann Arbor Science Publishers, Ann Arbor, MI 48106.

7. Chavan, V. V., and Mashelkar, R. A. (1980). Mixing of viscous Newtonian and non-Newtonian fluids, in *Advances in Transport Processes*, Vol. 1, (Ed. A. S. Mujumdar), Wiley Eastern Limited, New Delhi 110002, India.

8. Bird R. B., Armstrong, R. C., and Hassager, O. (1977). *Dynamics of Polymeric Liquids*, Vol. 1, Wiley, New York.

9. Middleman, S. (1977). *Fundamentals of Polymer Processing*, McGraw Hill, New York.

10. Carreau, P. J., De Kee, D., and Daroux, M. (1979). An analysis of the viscous behaviour of polymeric solutions, *Can. J. Chem. Eng.*, **57**, 135.

11. White, J. L., and Metzner, A. B. (1963). Development of constitutive equations for polymeric melts and solutions, *J. Appl. Polym. Sci.*, **7**, 1867.

12. Wichterle, K., Kadlec, M., Zak, L., and Mitschka, P. (1985). Shear rates on turbine impeller blades, *Chem. Eng. Commun.*, to be published.

13. Nienow, A. W., Wisdom, D. J., Solomon, J., Machon, V., and Vlcek, J. (1983). The effect of rheological complexities on power consumption in an aerated agitated vessel, *Chem. Eng. Commun.*, **19**, 273.

14. Peters, D. C., and Smith, J. M. (1969). Mixing in anchor agitated vessels, *Can. J. Chem. Eng.*, **47**, 268.

15. Feldkamp, K. (1962/1963). Untersuchung der Sekundarstroemung Nicht-Newtonscher Fluessigkeiten um rotierende Koerper. Diplomarbeit (Thesis), Technische Hochschule Braunschweig.

16. Uemura, T., O'Shima, E., and Inoue, M. (1967). Weissenberg effect in a high viscosity solution in a stirred tank, *Kagaku Kogaku*, **5**, 142.

17. Kelkar, J. V., Mashelkar, R. A., and Ulbrecht, J. J. (1972). On the rotational viscoelastic flows around simple bodies and agitators, *Trans. Inst. Chem. Eng.*, **50**, 343.

18. Kelkar, J. V., Mashelkar, R. A., and Ulbrecht, J. J. (1973). A rotating sphere viscometer. *J. Appl. Polym. Sci.*, **17**, 3069.

19. Walters, K., and Savins, J. G. (1965). Rotating sphere elastoviscometer, *Trans. Soc. Rheol.*, **9**, 407.

20. Ide, Y., and White, J. L. (1974). Rheological phenomena in polymerization reactors. Rheological properties and flow patterns around agitators in polystyrene–styrene solutions, *J. Appl. Polym. Sci.*, **18**, 2997.

21. Chavan, V. V., Arumugam, M. and Ulbrecht, J. J. (1975). On the influence of liquid elasticity in a vessel agitated by a combined ribbon screw impeller, *AIChE J.*, **21**, 613.

22. Carreau, P. J., Patterson, I., and Yap, C. Y. (1976). Mixing of viscoelastic fluids with helical-ribbon agitators—I—mixing time and flow patterns, *Can. J. Chem. Eng.*, **54**, 135.

23. Fruman, D. H., and Tulin, M. P. (1966). Diffusion of a tangential drag-reducing polymer injection on a flat plate at high Reynolds numbers, *J. Ship Res.*, **20**, 3.

24. Ford, D. E., Mashelkar, R. A., and Ulbrecht, J. J. (1972). Mixing times in Newtonian and non-Newtonian liquids, *Process Technol. Intern.*, (formerly *Brit. Chem. Eng.*) **17**, 803.

25. Coyle, C. K., Hirschland, H. E., Michel, B. J., and Oldshue, J. Y. (1970). Mixing of viscous liquids, *AIChE J.*, **16**, 903.

26. Chavan, V. V., and Ulbrecht, J. J. (1973). Power correlations for close-clearance helical impellers in non-Newtonian liquids, *Ind. Eng. Chem., Proc. Des. Dev.* **12**, 472; (1974). *Corrigenda* **13**, 309.

27. Chavan, V. V., Ford, D. E. and Arumugam, M. (1975). Influence of fluid rheology on circulation, mixing and blending, *Can. J. Chem. Eng.*, **53**, 62B.

28. Ford, D. E., and Ulbrecht, J. J. (1976). Influence of rheological properties of

polymer solutions on mixing and circulation times, *Ind. Eng. Chem., Proc. Des. Dev.*, **15**, 326.

29. Guerin, P., Carreau, P. J., Patterson, I., and Paris, J. (1984). Characterization of helical and screw agitators by circulation times, *Can. J. Chem. Eng.*, **62**, 301.

30. Ford, D. E., and Ulbrecht, J. J. (1975). Blending of polymer solutions with different rheological properties, *AIChE J.*, **21**, 1230.

31. Norwood, K. W., and Metzner, A. B. (1960). Flow patterns and mixing rates in agitated vessels, *AIChE J.*, **6**, 432.

32. Godleski, E. S., and Smith, J. C. (1962). Power requirements and blend time in the agitation of pseudoplastic fluids, *AIChE J.*, **8**, 617.

33. Wichterle, K., and Wein, O. (1981). *Int. Chem. Eng.*, **21**, 116.

34. Solomon, J., Elson, T. P., Nienow, A. W., and Pace, G. W. (1981). Cavern sizes in agitated fluids with a yield stress, *Chem. Eng. Commun.*, **11**, 143.

35. Giesekus, H. (1965). Secondary flow phenomena in general viscoelastic fluids, *Proc. 4th Int. Congr. Rheol.*, (Providence, Rhode Island, 1963), Vol. 1, 249; (1965). See also *Rheol. Acta*, **4**, 85.

36. Magnusson, K. (1952). Agitators for viscous liquids and two-phase systems, *IVA*, **23**(2), 86.

37. Ducla, J. M., Desplanches, H., and Chevalier, J. J. (1983). Effective viscosity of non-Newtonian fluids in a mechanically stirred tank, *Chem. Eng. Commun.* **21**, 29.

38. Van't Riet, K., and Smith, J. M. (1973). The behaviour of gas-liquid mixtures near Rushton turbine blades, *Chem. Eng. Sci.* **28**, 1031.

39. Calderbank, P. H., and Moo-Young, M. B. (1961). The power characteristics of agitators for the mixing of Newtonian and non-Newtonian fluids, *Trans. Inst. Chem. Eng.*, **39**, 337; (1962). *Corrigenda*, **40**, opposite p.1.

40. Skelland, A. H. P. (1967). *Non-Newtonian Flow and Heat Transfer*, Wiley, New York.

41. Chavan, V. V., and Ulbrecht, J. J. (1972a). Power consumption for mixing of inelastic non-Newtonian fluids by helical screw agitators, *Trans. Inst. Chem. Eng.*, **50**, 147.

42. Chavan, V. V., and Ulbrecht, J. J. (1972b). Power correlation for helical ribbon impellers in inelastic non-Newtonian liquids, *Chem. Eng. J.*, **3**, 308.

43. Chavan, V. V., and Ulbrecht, J. J. (1973c). Internal circulation in vessels agitated by screw impellers, *Chem. Eng. J.*, **6**, 213.

44. Bourne, J. R., and Buttler, H. (1969). Power consumption of helical ribbon imprellers in viscous liquids, *Trans. Inst. Chem. Eng.*, **47**, 263.

45. Chavan, V. V., and Ulbrecht, J. J. (1973b). Power correlation for off-centred helical screw impellers in highly viscous Newtonian and non-Newtonian liquids, *Trans. Inst. Chem. Eng.*, **51**, 349.

46. Patterson, W. I., Carreau, P. J., and Yap, C. Y. (1979). Mixing with helical ribbon agitators: part II—Newtonian fluids, *AIChE J.*, **25**, 508.

47. Yap, Y. C., Patterson, W. I., and Carreau, P. J. (1979). Mixing with helical ribbon agitators, part III—Non-Newtonian liquids, *AIChE J.*, **25**, 516.

48. Prokopec, L., and Ulbrecht, J. J. (1970). Power characteristics of a helical screw impeller with a draught tube, *Chemie-Ingenieur-Technik* (in German), **42**(8), 530.

49. Carreau, P. J. (1984). On the effective rate of shear in mixing of non-Newtonian liquids, *Chem. Eng. Commun.*, submitted.

50. Nagata, S. (1975). *Mixing, Principles and Applications*, Halstead, New York.

51. Bourne, J. R., Buerli, M. and Regenass, W. (1981). Power and heat transfer to agitated suspensions: use of heat flow calorimetry, *Chem. Eng. Sci.*, **36**, 782.

52. Bourne, J. R., Buerli, M., and Regenass, W. (1981). Heat transfer and power measurements in stirred tanks using heat flow calorimetry, *Chem. Eng. Sci.*, **36**, 347.

53. Deslouis, C., and Tribollet, B. (1975). Transport de matiere dans des solutions concentrees de polyoxyethylene a l'aide d'une electrode a disque tournant, *J. Chimie Physique*, **72**, 224.

54. Prud'homme, R. K., and Shaqfeh, E. (1984). Effect of elasticity on mixing torque requirements for Rushton turbines, *AIChE J.*, **30**, 485.

55. Mashelkar, R. A., Kale, D. D., and Ulbrecht, J. J. (1975). Rotational flows of non-Newtonian fluids, part 2. Supression of torque on agitators, *Trans. Inst. Chem. Eng.*, **53**, 150.

56. Ranade, V. R., and Ulbrecht, J. J. (1977). Gas dispersion in agitated viscous inelastic and viscoelastic liquids, *Proc. 2nd European Conf. on Mixing* (Cambridge), BHRA.

57. Quraishi, A. Q., Mashelkar, R. A., and Ulbrecht, J. J. (1976). Torque suppression in mechanically stirred liquids and multiphase liquid systems, *J. Non-Newt. Fluid Mech.*, **1**, 223.

58. Ranade, V. R., and Ulbrecht, J. J. (1978). Influence of polymer additives on the gas-liquid mass transfer in stirred tanks, *AIChE J.*, **24**, 796.

59. Bourne, J. R., Knoepfli, W., and Riesen, R. (1979). Batch and continuous blending of Newtonian fluids using helical-ribbon impellers, *Proc. 3rd European Conf. on Mixing*, BHRA Fluid Engineering, Cranfield, Bedford, England, Vol. 1, paper Al.

60. Hoogendoorn, C. J., and Den Hartog, A. P. (1967). Model studies on mixers in the viscous flow region, *Chem. Eng. Sci.*, **22**, 1689.

61. Johnson, R. I. (1967). Batch mixing of viscous liquids, *Ind. Eng. Chem. Proc. Des. Dev.*, **6**, 340.

62. Murakami, Y., Hirose, T., and O'Shima, M. (1980). Mixing with up and down impeller, *Chem. Eng. Prog.*, **76**(5), 78.

Chapter 5

Dispersion of Gases in Liquids:
The Hydrodynamics of Gas
Dispersion in Low Viscosity Liquids

JOHN M. SMITH

*Laboratorium voor Fysische Technologie, Technische Hogeschool Delft,
2600 AA Delft, The Netherlands*

1. INTRODUCTION

The dispersion of gases in liquids is one of the most commonly
applied processes of chemical engineering in general and of reaction
engineering in particular. It forms the basis of the vast majority of
mass transfer operations where the creation of adequate contact
area between the phases is a prerequisite of good design.

Many separation processes operate by generating this interfacial
area as a result of the flow through static packings or baffle systems.
In operations where mass transfer is gas film controlled, it is not
possible to improve greatly on this. However, in many reaction and
some absorption processes, mass transfer is liquid film controlled,
and in these circumstances the possibilities of affecting both contact
area and liquid film transfer coefficients are attractive. This
review will concentrate on the classical stirred tank geometry and its
operation. Attention is drawn to recent review articles by Midoux
and Charpentier[1], Brauer[2] and Schugerl[3], which deal with the
hydrodynamics and performance of a number of systems of practical
interest, while mass transfer has been considered in the reviews of
van't Riet[4] and Charpentier[5].

In equipment of given geometry, there are three independent
variables that are available to control performance: scale, stirrer

139

speed, and gas input rate. The process result that is desired may be dependent on the power demand, mass transfer coefficient, contact area, bubble size distribution, residence time distribution, and so on. The hydrodynamics of the impeller play a crucial role in the whole. Power consumption will be determined almost entirely by the local conditions in that region, and the other factors will be controlled by the relationships between the conditions near the impeller and those in the rest of the tank, including the very important contribution that the internal recirculation of gas makes to the impeller loading and performance.

1.1. Dispersion and Coalescence

The local bubble size distribution in a given gas–liquid mixture is developed as a result of a balance between the complementary processes of dispersion and coalescence. Breakup of a gas bubble must be preceded by distortion from the stable spheroidal equilibrium shape. The role of the various flow fields that can lead to breakup was detailed by Hinze[6]. Discussion of the fundamentals of turbulence theory and its application to mixing and reactor design will be found in the sections of this book which have been written by Brodkey and Patterson.

It is almost self-evident that the highly turbulent shear field near an impeller might disintegrate even those bubbles which are near the mean bubble size, while the more gentle conditions that prevail in the outer regions of the tank might permit similar bubbles to coalesce. Extensive data on the turbulence and velocity fields in the tank as a whole have been reported by Cutter[7] and Nagata[8].

Until a few years ago it was the almost universally accepted view that gas dispersion in a vessel agitated with a radial flow impeller occurred mainly in the turbulent discharge from the agitator, whilst coalescence was limited to that which occurred in the quieter regions.

More recently, it has become evident that another important mechanism is operative. The flow field around the impeller can induce considerable coalescence of gas bubbles before they experience subsequent redispersion in the outflowing liquid. In any event it seems unlikely that the larger bubbles are disintegrated as a direct result of interaction with random turbulence. Furthermore, it is clear that the complex three-dimensional flow that is generated around a rotating agitator in a single phase system is modified considerably when large amounts of gas are present. When the gas content is limited, the fundamental nature of the single phase flow field remains relevant. The ungassed hydrodynamics are therefore of interest and will be considered in a later section of this chapter.

1.2. Agitator Types

Agitators used for gas dispersion, at least in low viscosity systems, generally have small, high speed impellers producing either axial or radial flows. The most widely used is the radial flow turbine with four or six flat blades mounted on a disc (Fig. 1). This pattern was used by Rushton et al.[9] as a standard design. As well as providing the essential mechanical strength to the impeller assembly, the disc of a Rushton turbine has the advantage of limiting the short-circuiting of gas along the drive shaft. This can be a particular problem with multiblade paddles in large or unbaffled tanks when the centrifugal force field of the impeller tends to concentrate gas in the centre of the local forced vortex (Fig. 2).

This advantage of disc mounting highlights an important fact. The flow of multiphase systems adds a significant extra facet to behaviour of simple fluids; density differences can be ignored only with caution, and phase segregation may be encountered in those regions where dispersion should be occurring.

For reactors in which suspension of solids has to be maintained at the same time that a gas must be dispersed, there is some advantage in using a pitched blade agitator. Fermentors are often designed with large blade agitators of relatively limited cross sectional area. An example is the SuperMIG from EKATO[10] (Fig. 3).

Heavier solids may require stirrers with a larger pumping

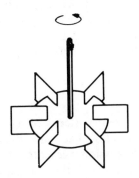

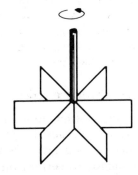

Figure 1. A six blade (Rushton) disc turbine.

Figure 2 A multiblade paddle, hub mounted turbine.

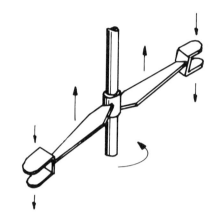

Figure 3   The SuperMIG (Ekato) impeller.

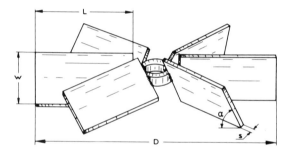

Figure 4   A pitched blade turbine.

capacity, and pitched blade turbines are being used increasingly for this duty (Fig. 4). The more idealised form of a marine propeller has limitations of both mass and cost for large scale applications. Hub-mounted radial flow flat blade turbines are in fact an intermediate form between the pitched blade and Rushton patterns, though they lack the advantages of either.

1.3.   Geometric Variables

Apart from consideration of grossly different patterns of agitators, there is a multitude of geometric variables that can affect the performance of a given design. The discussion presented here is largely limited to performance in equipment with the configuration that has become known as the standard geometry, and which is shown in Fig. 5, together with the notation used.

Most published information on mixing vessels relates to tanks

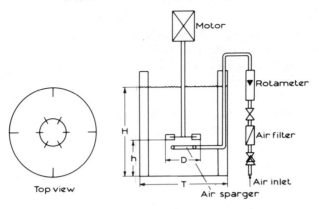

Figure 5   The standard tank geometry.

which have flat bottoms. These are normally filled to a depth H equal to the tank diameter T. For gas dispersion applications the impeller usually has a diameter D between 0.25 T and 0.5 T, and is mounted between D and $H/2$ above the base, though for solids suspension sometimes even smaller off-bottom clearances are used. The tank is normally fitted with four full length vertical baffles of 0.1 T width, mounted against the walls. For some applications where special care is needed to avoid dead zones, a small clearance may be left between the baffles and the tank walls and bottom.

Major deviations from this geometry are found in industrial practice. Tanks usually have dished or conical ends, and much taller vessels with multiple impellers are used for some more difficult gas absorption applications, for example, in fermentor designs. There is no agreed optimum (or even reference) design for the gas inlet or its position. It seems that, at least for Rushton turbines, sparger rings smaller than the impeller in diameter and mounted within one impeller diameter of the disc plane perform satisfactorily, and the details of their installation are not usually critical. That this does not necessarily hold for other impeller designs will become clear when gassed pitched blade turbines are discussed.

2.   GENERAL CONCEPTS

2.1.   Primary and Secondary Flows

A rotating impeller, whether designed to produce axial or radial flows, first induces a rotation in the fluid (a primary flow) which is converted by centrifugal and shear forces into the desired bulk motion. It is usual to use baffles to break up the primary rotation and to ensure efficient randomisation of the flow.

The distinction which has been made here, between the conditions near the impeller and those in the rest of the vessel, applies not only to the fluid motion, but to all the aspects of stirred tank operation. We will see that gas dispersion is controlled by the immediate impeller region, whereas coalescence processes can be expected to be more important in the quieter, more remote parts of the system.

Similarly, such variables as gas holdup distribution and recirculation, or film coefficients for mass or heat transfer, are determined by the interplay of conditions both near and far from the agitator. However, the impeller hydrodynamics control two very important variables directly: the power demanded by the system to maintain the agitator rotation at its given value and the directly induced fluid motion that is evident in the pumping capacity of the impeller. In a gas–liquid system the characteristics of the freshly produced two-phase mixture in this discharge stream are also largely determined by the local conditions at that point.

2.2. Power Consumption: Single Phase Flows

Since the interaction between the impeller and the system is most easily examined in terms of the shaft power demand, this provides a good starting point for the consideration of impeller operation. It should be emphasised that studies of power consumption are not only useful in terms of providing necessary design specifications but also give direct insight into the conditions in the impeller locality.

The couple required to maintain the rotation of a disc in water was first investigated by Thomson[11]. He found that the power varied as the third power of the disc speed and the fifth power of the diameter: it was evidently a fully developed turbulent system. He discovered also that the power was affected by the clearance between the disc and its housing (his interest was in the fundamentals of the design of centrifugal pumps) but no publication of follow-up work is known to the present author.

The power demands of standard impellers follow the curve shown in Fig. 6, which has been taken from Dickey and Fenic[12]. This shows the graph of power number

$$\text{Po} = P/N^3 \cdot D^5 \cdot \rho \tag{1}$$

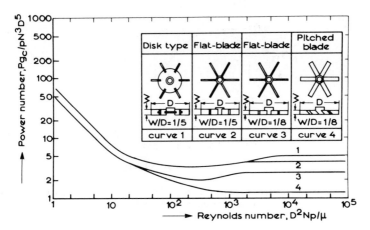

Figure 6   Power number—Reynolds number relationship for turbines and propellors
(after Dickey and Fenic[12]).

against the stirrer-based Reynolds number

$$Re = N \cdot D^2 \rho/\mu \, . \tag{2}$$

For all agitators the laminar flow regime is characterised by a linear decrease in the power number with Reynolds number, while in a fully turbulent system, as Thomson found, the power number is relatively constant. When operating in a gas–liquid mixture, a given impeller almost invariably has a smaller power demand than when driven at the same speed in the absence of gas. This is partly a consequence of the reduced density of the pumped mixture, and partly a result of the streamlining effect produced by stable ventilated cavities that develop behind the stirrer blades. The change in power demand between single and two phase conditions provides a very useful basis for comparison. This is reflected in important process variables such as gas hold up and mass transfer performance.

The form of the relationship between the power number and the gas flow number,

$$Fl = Q_g/N \cdot D^3 \, , \tag{3}$$

usually follows the pattern shown in Fig. 7, which actually applies for a six-blade-disc turbine operated at constant speed. The shape of the curve is related to the modification of the form of the ventilated cavities as the balance between the supplied gas volume rate and the

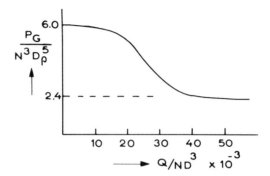

Figure 7   Power number gas flow number relationship for a disc turbine operating at constant speed in the turbulent regime.

nominal liquid pumping rate changes. The power demand characteristics of various geometries with different degrees of gas loading are dealt with in greater detail in the relevant sections of this chapter.

2.3. Pumping Capacity and Liquid Circulation

The pumping capacity of an impeller, that is, the volume of fluid discharged through it per unit time, is proportional to the product of the tip speed of the agitator and its discharge area, that is, in total to the third power of the diameter. The circulation and mixing within the tank volume as a whole is not of primary concern here, but it is important to realise that small bubbles can be recirculated to the impeller region as a result of the bulk liquid flow. In this way the actual gas loading on the impeller may be significantly greater than the gas rate leaving the sparger.

The discharge from an impeller entrains more liquid from the regions on either side, so that not all the recirculated liquid, nor the associated gas either for that matter, necessarily passes through the swept volume of the agitator, (Fig. 8). This makes direct evaluation of the pumping capacity of an impeller difficult, and the determination of the performance in two phase flow systems even more so. These remarks serve to highlight an area of ignorance that still remains. Before gas–liquid systems can be said to be fully understood, this information on liquid pumping, entrainment, and recirculation must be forthcoming.

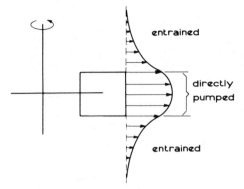

Figure 8   Direct and entrained outflow velocity profiles from a disc turbine.

2.4.   Flow Regimes in Aerated Stirred Tanks

It is useful to consider the various flow patterns that can be encountered in aerated stirred tanks at this stage. We will limit consideration to tanks operating with a single dispersing impeller. There are differences in principle between the successive flow regimes established by the various types of impeller. Radial and axial flow devices generate quite different flows both locally and throughout the tank as a whole.

The gas distribution produced by a radial flow impeller at a given supply rate depends on the stirrer speed. At low stirrer speeds, the buoyancy induced flows dominate the circulation. The impeller is ineffectual and can be said to be flooded (Fig. 9a). At somewhat higher speeds the radial pumping action is sufficient to load the impeller and to produce a dispersing action (Fig. 9b). The liquid velocities are limited and the gas bubbles rise through the upper part of the tank with little further disturbance, finally escaping from the system at the free surface.

The increase in the volume of liquid circulated by a further rise in the impeller speed leads to downward velocities in some regions of the liquid which are sufficient to prevent bubbles rising. Recirculation of gas to the impeller can result. Whether this first occurs above or below the impeller plane will depend primarily on the D/T ratio. Figures 9c and 9d show this situation, which is discussed further in the section dealing with high gas loadings and flooding criteria. It is in this range that the local conditions begin to interact with the flow in the tank as a whole. Only at still higher speeds is the liquid flow fast enough to ensure that the gas is distributed more or

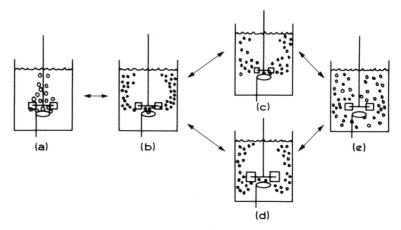

Figure 9   Successive flow regimes for a disc turbine with decreasing gas flow number.

less uniformly throughout the liquid volume (Fig. 9e). Although a similar picture is provided by downward pumping axial flow impellers (Fig. 10), this is modified by the fact that gas is released into a liquid flow which tends to prevent it reaching the impeller. Figure 10a shows the situation at low stirrer speeds. Again buoyancy forces dominate the bulk circulation; the impeller is flooded, dispersion is ineffectual and indeed the mean liquid flow through the impeller may be opposed to the nominal pumping direction. When the impeller rotates more quickly, a point is reached when the gas begins to be dispersed and to be conveyed downwards (Fig. 10b). Under these conditions the impeller can be said to be directly loaded with gas from the sparger. In equipment with a reasonably small D/T ratio it is quite possible for the liquid returning to the impeller from above to have insufficient velocity to capture gas and

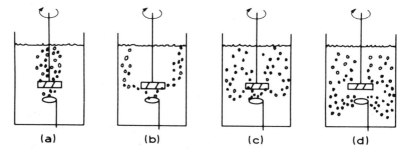

Figure 10   Flow regimes for an axial flow turbine, with decreasing gas flow number.

recirculate it. At higher stirrer speeds, when the liquid velocities are of the same magnitude as the free rise velocities of individual bubbles, about 0.3 m/s, such recirculation may occur. However the discharge from an axial flow turbine is normally divergent (leading to Chapman et al.[13], calling them "mixed flow" impellers), and there may even be a small region of upflow located centrally below it. This effect can be great enough to allow gas released below the impeller to reach it and load it directly. This condition is shown in Fig. 10c.

A further increase in the stirrer speed changes the picture. The downflow can be rapid enough to prevent direct access of the introduced gas to the agitator. Under these circumstances the liquid velocities are sufficient to ensure recirculation of gas to the impeller plane from above, and the impeller can be said to be indirectly loaded (Fig. 10d). It will be evident immediately that not only is the D/T ratio relevant, but that the exact details of the gas supply arrangements, notably the sparger ring diameter and the distance between it and the impeller plane, are important.

3. SINGLE PHASE SYSTEMS

3.1. Visualization

Much of the detail of the flow field in the immediate locality of an impeller has been established as a result of tomographic observations, that is, measurements made, or at least recorded, in a coordinate system which is based on the agitator itself, and rotates with it. Tomographic recordings of the flows within various types of multiblade impeller were published by Takeda and Hoshino[14], who made streak photographs of the fluid motion using single frame cameras that rotated in synchronism with the impeller shaft. More extensive details of the flow field can be obtained from stroboscopic or film techniques. One example of reporting in this way was provided by Beckner[15], who determined the flow field produced by an anchor agitator by manually transcribing frame by frame data from a cine film taken with a stationary camera. The relatively slow rotations needed for anchor studies encouraged Peters and Smith[16] to reduce the labour involved in this method by mounting a film camera on an extension to the mixer shaft. However, even for these not too arduous conditions, there were mechanical difficulties with conventional film cameras. These can be avoided by using television cameras, as described by Smith[17] (Fig. 11). TV cameras can operate

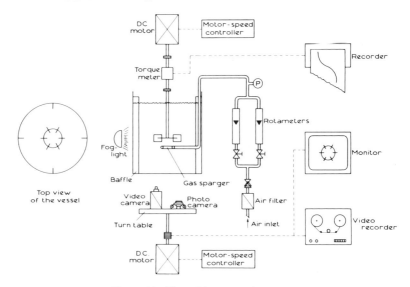

Figure 11   Turntable mounted cameras.

under much more severe conditions of dynamic loading, and have been extensively used in the studies of van't Riet[18] and War-moeskerken[19].

3.2.   The Trailing Vortex System

3.2.1.   The Velocity Field

Detailed measurements of the velocity field produced by Rushton turbines in the turbulent single phase regime were made by van't Riet[18]. The outflow is dominated by a trailing vortex pair behind each blade of the turbine (Fig. 12). Both video and single frame photographs were used to determine the velocities associated with

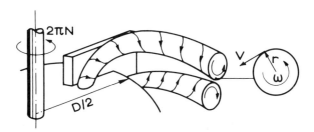

Figure 12   The trailing vortex pair behind a disc turbine blade.

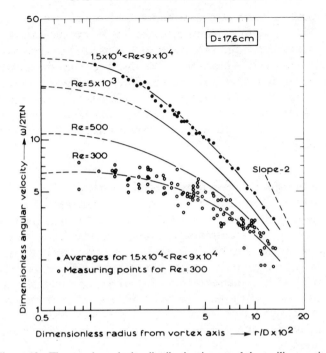

Figure 13   The angular velocity distribution in one of the trailing vortices.

one of these vortices. The vortex can be approximated by an idealised line vortex within which the liquid follows a helical path at a constant radius from the vortex axis. On this basis it is possible to estimate the magnitudes of the rotational velocity components and

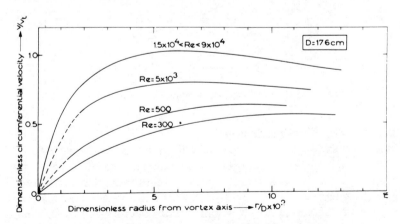

Figure 14   The circumferential velocity component in a trailing vortex.

the associated shear rate distribution and centrifugal acceleration fields. The conditions relating to the blade tip provide a useful scaling variable for the properties investigated. The dimensionless angular velocity as a function of the distance from the axis of the idealised vortex is shown in Fig. 13. When translated into a diminsionless circumferential velocity component this has the form shown in Fig. 14. The uniformity of this velocity with radius is remarkable, particularly for the high Reynolds number regime that is relevant for gas dispersion.

3.2.2. Centrifugal Forces in the Trailing Vortices

In a water-like low viscosity system the circumferential velocity is approximately the same as the blade tip velocity to within a very few millimeters of the centre of rotation. This leads to very large centrifugal forces. Figure 15 shows the general magnitude of these forces estimated from an idealised fit of the measured data. For an impeller of 0.176 m diameter operating with a tip speed of 2 m/s the maximum value corresponds to about 500 g near the vortex core. It is self-evident that centrifugal forces of this magnitude have a very great influence on the pressure distribution and possible phase separation.

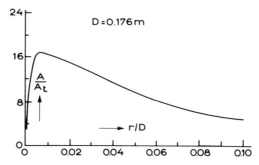

Figure 15 The centrifugal acceleration in an idealized vortex.

3.2.3. Shear Rate Distribution

Figure 16 shows the shear rate distribution that has been derived from the same velocity data. The maximum shear rates occur not far from the vortex axis, and are about five times those that control the average shear rate as derived by Metzner and Otto[20] on the basis of the power demand in non-Newtonian fluids operating in the laminar flow regime.

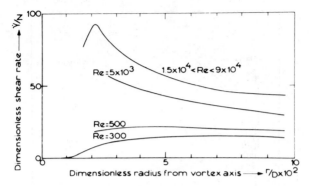

Figure 16   The shear rate distribution in an idealized vortex.

3.2.4.   The Underpressure

The pressure distribution behind a turbine blade was determined in a single phase system by van't Riet[18]. The pressure sensor was mounted on the stirrer shaft and rotated with the agitator. This allowed exploration of the pressure field within and immediately adjacent to the impeller. Results for two different impellers are shown in Figs. 17 and 18. The isobars are drawn, again using the

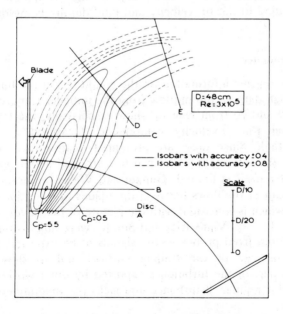

Figure 17   Downpressure behind an impeller of 0.48 m diameter.

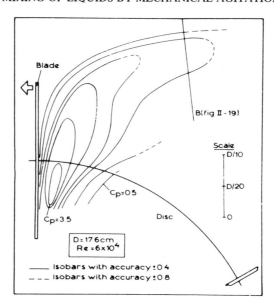

Figure 18   Downpressure behind blades of a 0.176 m diameter impeller.

blade tip conditions as a basis, that is, the stagnation pressure corresponding to the tip velocity and fluid density being defined as +1.

3.3. Turbulence

There is extensive information on the conditions in the bulk of the fluid, though data on the outflow from the impeller is more limited. Figures 19 and 20 from Nagata[8] show respectively, the lateral and longitudinal r.m.s. velocity components measured in a 0.3 m diameter tank. Since these are presented relative to a stationary tank wall, the information that can be extracted about the flow in the impeller itself is limited. Gunkel and Weber[21] presented more detailed data of the flows between the blades of a turbine impeller (Fig. 21), which is consistent with the picture of the rapidly spinning vortices of Fig. 12. Van't Riet and Smith[22] were able to relate data obtained from fixed probes to the signals to be expected from the repeated passage of the trailing vortices and to show that a significant part of the turbulence reported by other workers is the result of this regular disturbance and lacks the randomness of true turbulence.

Since the circumferential velocity is of the same order as the tip

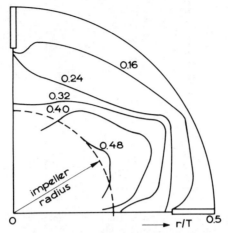

Figure 19   Lateral r.m.s. velocities in a stirred tank (after Nagata[8]).

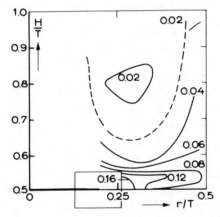

Figure 20   Longitudinal velocities in the impeller plane (after Nagata[8]).

velocity, and about one fifth of the outflowing fluid is associated with the trailing vortices, an average r.m.s. velocity of around $0.2 \, v_t$ is to be expected. Although Nagata's data shows very different lateral and longitudinal r.m.s. components at about 0.1 and $0.25 \, v_t$ they do lie reasonably either side of van't Riet's results.

Developments in laser-doppler anemometry will lead to rapid advances in the knowledge of flow and turbulence developed by turbines. This is illustrated by a recent paper by Popiolek et al.[23]. These preliminary results suggest rather lower velocities in the trailing vortices than those reported by van't Riet.

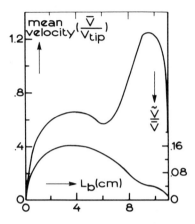

Figure 21 Local outflow and r.m.s. components in mean velocity distribution between blades (after Gunkel and Weber[21]).

3.4. Implications for Two-phase Systems in General

The centrifugal force fields and turbulence in the convected outflow from the impeller exert complementary actions in a two-phase system. The trailing vortices will tend to fling out any heavier material which may be present, particles will impact the rear surface of the impeller and it is probable that this represents a significant source of secondary nucleation in crystallisers. It is likely that any particle erosion damage would be concentrated near the disc plane on the rear of the blade, with some wear on the front faces of the impeller blades near the upper and lower edges (Fig. 22). The low pressures can cause cavitation to occur with high speed impellers particularly in fluids which are near boiling point or which have a high vapour pressure of dissolved components. The vapour cavities

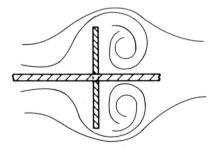

Figure 22   Flow paths over a turbine blade in single phase operation.

would form in fluid that is moving rapidly away from the impeller, and their collapse would in any case not occur in contact with the blades. Cavitation erosion is not to be expected therefore.

Liquid–liquid operations provide a relevant contrast to the gas–liquid processes that are considered here. The differences that arise as a result of the much higher inertia associated with the dispersing phase in this instance are perhaps usually overshadowed by the consequences of the lack of compressibility. Liquid–liquid contacting operations must also be influenced by these intense centrifugal force fields. Almost all the published data on drop size distribution and mass transfer rates have been obtained for systems in which the dispersed phase has been the lighter of the two, and for which the mechanisms of drop coalescence and dispersion may not differ too much from the those relevant to gas–liquid systems. However, it should be expected that if the heavier phase is being dispersed, then quite different mechanisms would be important. A priori it is surprising that, almost without exception, correlations relating to liquid–liquid dispersion involve some functions of density difference, without regard to which phase is being dispersed, though of course the major part of the available data has been obtained for dispersing a light organic system in water. Another conclusion of practical relevance that can be drawn from the single phase hydrodynamic studies is the clear advantage to be expected from feeding reagents of a fast reaction to locations just inside the vertical inner edges of disc turbine blades. The feedstocks will then pass through the trailing vortex zones of maximum shear and turbulence, and will be dispersed rapidly and efficiently. This procedure will be far more effective than the usual practice of feeding into the bulk outflow discharge stream, a location that ensures that the maximum possible time will elapse before the fluid experiences the most severe mixing conditions which exist in the vessel.

3.5. Pitched Blade Agitators

There has recently been considerable interest in the hydrodynamics of pitched blade turbines. The flow field differs from that produced by a Rushton turbine in that the circulation around a blade in axial flow produces a single tip vortex. The flow structure has been described in detail by Tatterson[24]. Figure 23 shows the main features. The stagnation line in the approaching flow is a little below and to the front of the leading edge of the blade. Near the shaft axis there is a rapid downflow with little rotation. A strongly spinning tip

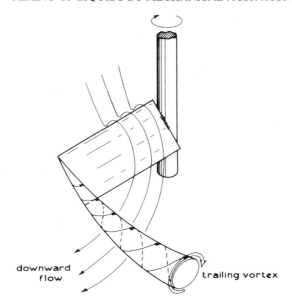

Figure 23   Flow induced by a pitched blade turbine (after Tatterson[24]).

vortex extends a considerable distance downstream. A very similar outflow is generated behind propellors designed for marine propulsion, and the associated cavitation phenomena are well known.

The flow produced in a stirred tank is very dependent on the impeller geometry, particularly the angle of attack of the blades, the distance from the tank bottom, and the precise shape of that bottom, whether it is flat, dished, or otherwise profiled. Details of the outflow from various axial flow turbines have been published by Fort[25]. Figure 24 shows typical results. However, the importance of the stirrer position on the overall flow pattern is of greater interest. Figure 25 shows the situation when an impeller is mounted near the tank bottom. Despite the nominal axial flow characteristics of the impeller, the induced fluid motion is strongly radial. An axial flow agitator mounted remote from the tank bottom can also generate a small region of reversed flow centrally below the shaft. The primary rotation of the fluid and the effects of the drag on the bottom of the tank produce a secondary circulation. This results from the pressure gradients produced by the imbalance between the differing centrifugal forces at various distances from a fixed bottom in fluids which are rotating. The resulting motion over the bottom of the tank has a significant sweeping action from the outside towards

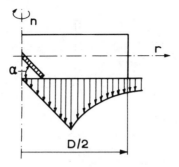

Figure 24 Velocity profile in outflow from a pitched blade turbine (after Fort[25]).

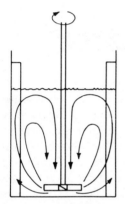

Figure 25 Flow field with an axial flow impeller mounted near the tank bottom.

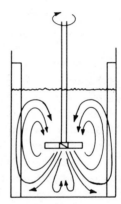

Figure 26 Flow field of an axial flow impeller mounted near the tank mid plane.

the centre. This effect is perhaps best known in the gathering together of sediment, for example, tea leaves, at the bottom of a vessel which has been stirred. Although this flow is weak, it may nevertheless determine the behaviour of gas introduced from a small centrally mounted sparger.

4. GAS–LIQUID SYSTEMS WITH LOW GAS FRACTIONS

4.1. Bubble Capture by Trailing Vortices

Any gas in the impeller region is directly influenced by the hydro-dynamic regime, and in turn it can modify that regime considerably. The large axial pressure gradients along the trailing vortices were

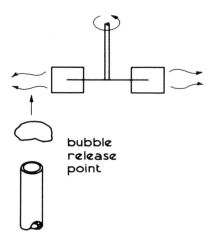

Figure 27   Experiment to study the behaviour of bubbles released into the turbulent outflow from a radial turbine.

confirmed spectacularly by the observations reported by van't Riet and Smith[26]. The experiment is sketched in Fig. 27. Bubbles were released so as to rise into the discharge from a Rushton turbine, with the objective of studying the breakup of large bubbles in the highly turbulent outflow. The rotating TV camera equipment was used. The sequence of photographs, Figs. 28, 29, 30, and 31 show that the gas is drawn upstream against the radial outflow. The liquid is moving with a velocity far greater than the free rise velocity of bubbles in stagnant liquid, and the gas bubble first moves up

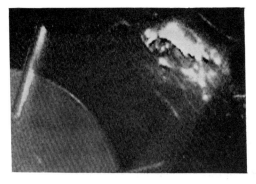

Figure 28 $t = 0$.

Figures 28–31 Sequence showing the capture and subsequent disintegration of a large bubble released into the outflow from a Rushton turbine.

Figure 29 $t = 0.085$ s.

Figure 30 $t = 0.165$ s.

Figure 31. $t = 0.205$ s.

alongside the blade. Only subsequently does the bubble lose gas by breakaway from the elongated tail into the rapidly spinning liquid, by a mechanism that may be related to the disintegration of a Rayleigh jet. The underlying mechanism of breakup of large bubbles that enter the radial outflow from a turbine is obviously quite different from that which has been conventionally imagined. Dispersion is almost inevitably coupled with coalescence and the immediate agitator region has an important role to play in this.

4.2. Vortex Cavities

The capture process described above, in which gas is drawn against the prevailing liquid flow to form a cavity behind a turbine blade, is not normally the main mechanism acting in an aerated stirred tank. Certainly, the low pressure region and the strong centrifugal forces influence the flow paths taken by the gas bubbles, but in general gas approaches from over or under the blades, as can be imagined from the flow lines drawn in Fig. 22.

The bubbles form stable spinning cavities behind the blades (Fig. 32). The general appearance suggests that the cavities are filled with foam, though there is a large pressure gradient which will cause the bubbles to expand as they are drawn together into the vortex. These circumstances make coalescence very likely. The vortex cavities maintain a steady size as a result of a balance between the rate of accretion and the losses from the elongating tail. The volume of gas which is held in the cavities at any time depends on the stirrer speed and the total gas rate through the agitator region, including possibly

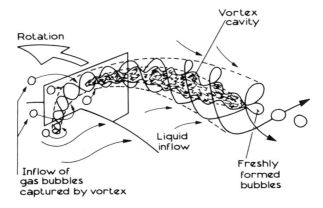

Figure 32   Gas filled vortex cavities.

a contribution from recirculation. A pair of vortex cavities usually forms behind each of the blades of a disc turbine. At low stirrer speeds, with little recirculation back to the agitator plane, it is possible to observe gas filling only the lower of the vortices of each pair, as it is captured as it leaves the sparger and undergoes just one cycle of coalescence and redispersion before escaping from the system at the free surface.

The gas, by occupying the vortex core, reduces the mass of fluid experiencing the extreme conditions there. The consequence of this is that there is a small reduction in the energy dissipation, and thus the power demand, and the highest shear rates to which the liquid is subjected is reduced. The impeller will also be pumping a mixture of somewhat reduced mean density, and this fact probably also contributes to the drop in power level.

Vortex cavities are characteristic of a regime in which the gas fraction is so small that there is no major modification in the single phase flow field. Such conditions only apply when the gas supply is limited, for example to that produced by air entrainment through the free surface. Such gas can only reach the impeller by a mechanism like that producing recirculation, which will be discussed further in the section dealing with high gas loadings.

The limited importance of entrainment through the free surface was shown by Matsamura et al.,[27] who found that even with very small gas throughputs, surface entrainment is only equivalent to about 5% of the superficial gas velocity (Fig. 33a). Matsamura et al. also demonstrated that the rate of entrainment is sharply reduced in saline solutions which exhibit reduced coalescence (Fig. 33b).

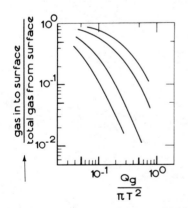
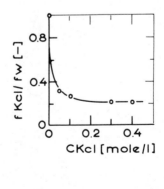

Figure 33 (a) Air entrainment rates through surface aeration. (b) Reduction in surface aeration in a saline solution, (Matsamura et al.[27]).

4.3. Clinging Cavities

If the gas flow number (Q/ND^3) of an impeller operating in the vortex cavity regime is increased, the cavities grow in both length and diameter. Since the natural location of the axis of the trailing vortex behind the blades does not shift position significantly, the spinning liquid has less and less room between the cavity and the rear of the blade. Eventually the rotation is sufficiently hindered that the gas extends right up to the blade and a different, transitional cavity form which is attached to the blade is established (Fig. 34). Clinging cavities are formed uniformly around the turbine, that is, each blade will be loaded with a merging pair of cavities, all of which will be of the same shape. The liquid flow associated with clinging cavities is still primarily radial, with gas dispersion taking place from the outer surface. There is still some residual rotation to be seen in the outflow, though this is much less than when vortex cavities are present.

The gas throughput has a greater influence on the size of clinging cavities than is the case with vortex cavities. This is probably a result of the reduced rotation in the two phase mass which leads to a less efficient dispersing action at the rear of the cavity. The steady growth of the clinging cavities presumably reduces the pumping capacity and possibly also the mean density of the pumped fluid. At the same time the effective drag coefficient of the blades is reduced as a consequence of the improved smoothness of the flow around the blades. One result of this is that the fall in power demand of the agitator steadily steepens with increasing gas flow number as long as the clinging cavity regime is maintained.

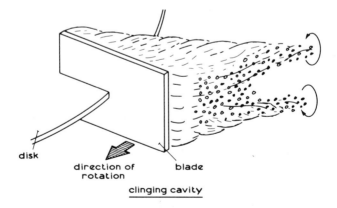

disk

direction of   blade
rotation

clinging cavity

Figure 34   A clinging cavity.

4.4. Power Consumption in the Vortex and Clinging Cavity Regimes

The power demand of an agitator is less in an aerated system than when it operates in liquid alone. Figure 7 has shown how the power number changes as the gas flow number is varied. The exact values at which these cavities change from one form to the other is controlled by the geometry of the system, including the clearance between the agitator and the bottom of the vessel, and by the extent of recirculation[28]. For the present it is sufficient to say that the cavities are of the clinging or vortex types only if the gas flow number is less than about 0.03, which, as can be seen from Fig. 35, is a point of inflection on the power curves. The initial drop from the ungassed power level is fairly linear for a standard six-blade-disc turbine, but steepens as the cavity form changes. This must reflect a decreasing effectiveness of the impeller in pumping the liquid. The cumulative effect is to lead to relative power levels of about half those in liquid alone. There does not seem to be a sharply defined point of transition from the vortex cavity to the clinging form, but the fact that the changes are gradual and progressive means that there is no overwhelming need to be able to predict the changeover.

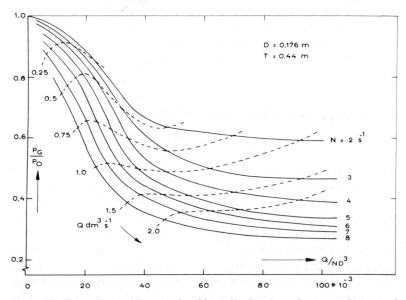

Figure 35 Power demand by gassed turbines showing dependence on stirrer speed and gas rate.

The significance of recirculation in determining the gas loading of the impeller was examined by van't Riet[18]. Experiments carried out in vessels of different sizes with baffles extending inwards, so that the clearances between the impeller and the baffles were the same, allowed comparison of the power demand under normal conditions with that when recirculation was avoided, making it possible to estimate the degree of internal recirculation of gas above the impeller.

The study also showed that the recirculation was dependent on the stirrer speed, and it follows that the true total gas loading on the agitator cannot be fully defined by the single dimensionless gas flow number, and this is the reason that the matrix of lines shown on Fig. 35 is necessary in order to describe the differing effects of the gassing rate and the stirrer speed. The importance of recirculation has been discussed by Nienow et al.[29] and Chapman et al.[13]. By plotting the power demand for curves in which the gas rate is held constant (the dashed lines in Fig. 35), these authors distinguish minima and maxima that they regard as reflecting the changes in the hydrodynamic conditions. This approach of varying the stirrer speed is of course the normal one used for studies in liquid–solid systems, though it does have the disadvantage in the present application that it is impossible to approach the $Fl = 0$ asymptote. However, the matrix of Fig. 35 is composed of interrelated curves and this point is of little consequence. The implications of the steep fall in the power curves obtained at constant stirrer speeds will perhaps become clearer from the discussion of performance at intermediate gas loadings.

4.5. Pitched Blade Stirrers at Low Gas Rates

Gas loading of a pitched blade turbine initially follows the pattern that might be expected in the light of our knowledge of what happens with Rushton turbines. At low gas rates the tip vortex captures gas in a very similar way. However, the circulation that goes right around the blade of an axial flow turbine has no counterpart with a disc mounted blade, and so the circulating eddy that might develop behind the upper rear edge of the blade is less intense with a pitched blade turbine. This results in a less efficient capture of gas by the trailing vortices and a pitched blade turbine can pump a bubbly mixture with less coalescence than a disc turbine can. Nevertheless the successive stages of vortex and clinging cavities are found and the power demand is affected. A power curve is

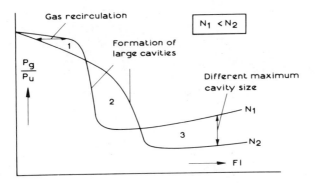

Figure 36   Typical power demand of a gassed pitched blade turbine.

shown in Fig. 36.[19] We are again concerned here with the initial part of this curve. It can be seen that once more there is a modest drop in the power consumption, though the fall is rather less than that with a Rushton turbine, though the situation with an axial flow impeller is more complex as a result of the additional importance of such details of the geometry of the equipment as the separation between the sparger and the impeller. The transition that leads to the very sudden drop in power level will be considered in the section dealing with intermediate gas loadings.

5.   HYDRODYNAMICS AT INTERMEDIATE GAS LOADINGS

5.1.   Large Cavities

It has been pointed out above that the transition between vortex and clinging cavities is gradual and has little consequence for the process. The sizes of these cavities are determined by the balance between stirrer speed and gas rate. As the gas flow number is increased to about 0.03, there is a sudden and fundamental change in the impeller hydrodynamics, as cavity form changes with the development of large ventilated cavities (Fig. 37). The liquid outflow near a large cavity is more tangential than near a clinging cavity. A large cavity is a single gas bubble with essentially smooth lateral surfaces which shrouds virtually the entire rear surface of the stirrer blade. Gas loss from the cavity in low viscosity fluids is associated with a turbulent breakup of the rear surface, with little or no evidence of the elongational spinning flow that is so efficient behind vortex cavities. The rather similar cavities formed in high viscosity

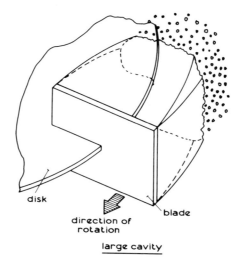

Figure 37   A large cavity.

fluids lose gas in a rather different way, an anvil shape develops and breakaway of fine gas bubbles occurs from the streaming tip. It has been established that several process parameters are influenced by this change in cavity form. The mass transfer factor, gas holdup and power demand curves all show different dependences on the operating variables depending on whether or not these large cavities are present. For this reason some effort has been directed towards establishing the circumstances that determine the conditions that lead to and maintain large cavities.

5.2. Observations of Large Cavity Development

Bruijn et al.[30], working with Rushton turbines of various sizes and directly observing the hydrodynamic regime within which the impellers were operating, related the steadily decreasing power demands of the agitator to the influence of large cavities. The observations were in error in so far as it was reported that large cavities could be formed one at a time until all six blades were streamlined. The authors now realise that they were misled by observations that were based on single frame photographs taken of equipment in which the incoming gas was probably not well distributed around the turbine. It was however reported correctly that in the regime when large cavities are just formed, then these are never to be found behind adjacent blades of a standard turbine. A more accurate technique was developed by Warmoeskerken[31] using a flexible vane with strain gauges which was mounted in the turbine outflow to deduce the flow conditions. Figure 38 shows the experimental arrangement in which the vibrations of the detector could be related to visual observations, and Figs. 39 and 40 give details of the construction of the vane and the instrumentation used with it respectively. The technique depends on the fact that there is a difference in the character of the liquid outflow depending on whether there is a cavity behind a given blade or not. In the absence of gas, the discharge from a six-blade-turbine is dominated by the blade passage frequency, and the strain gauges mounted on the flexible vane produce a signal which can be idealised to the form shown in Fig. 41. In practice the signal is very noisy, and interpretation is far easier if the frequency spectrum of the signal is used. An example is shown in Fig. 42. The blade passage frequency is seen as a single distinct peak in the spectrum. When a six-blade-disc

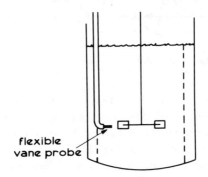

Figure 38   Vane detector arrangement.

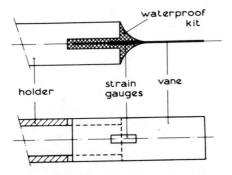

Figure 39   Vane construction.

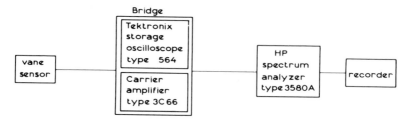

Figure 40   Circuitry for analysis of vane signal.

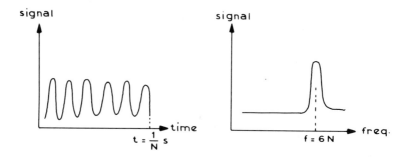

Figure 41   Principal vane deflections   Figure 42   Corresponding frequency
with no gas present.   spectrum in the absence of gas.

turbine has three large cavities, these are located behind alternate blades. The outflow signal then appears rather as is shown in Fig. 43, while the spectrum now shows two peaks, one at the blade passage frequency and one at half that, Fig. 44.

It was found that not only was this configuration of large cavities behind alternate impeller blades particularly stable, but that its

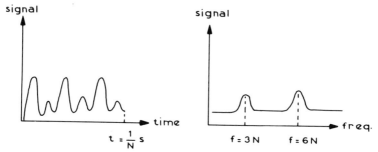

Figure 43   Vane deflections when three   Figure 44   Frequency spectrum cor-
large and three clinging cavities are   responding to the 3–3 state.
present.

initiation was very sharply defined. Moreover the blade loadings involving only one or two cavities are essentially transitional states, and cannot be maintained. The three cavity configuration is thus the first stable one to be reached as the gas flow number increases.

Figure 45 shows the relationship found for the onset of this configuration as determined by observation with the rotating TV camera and with the flexible vane detector. The agreement between the methods is clearly satisfactory. The straight line drawn in the figure corresponds to a gas flow number, Fl, of 0.030. This line divides the operating field between a regime with only clinging cavities present and that with large cavities. It can be seen that this transition occurs at the same value of the flow number in both flat and round bottomed tanks. Further tests have proved that even in non-coalescing systems such as saline solutions, there is a minimal change in this transition, the additional recirculation resulting from the much finer bubble dispersion only reducing the transitional flow number to 0.029.

These experiments were made in a tank of 0.44 m diameter, with an impeller of 0.4 T diameter. Further tests with a range of impellers in a given tank have shown that the transitional flow number

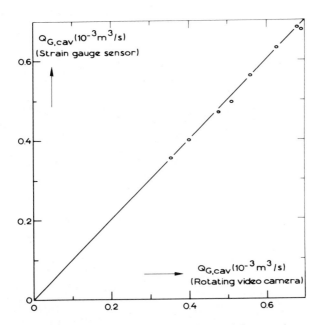

Figure 45 Comparison between visual observations and the vane detector results.

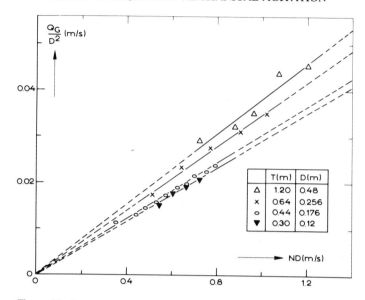

Figure 46   Detected cavity formation conditions for various geometries.

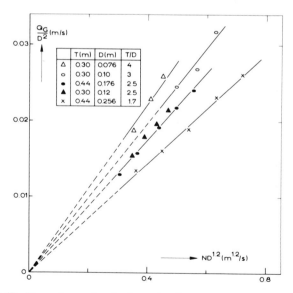

Figure 47   Dependence of cavity formation line on tank to diameter ratio.

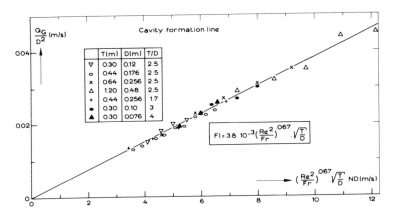

Figure 48   Scale independent cavity formation line.

depends on the D/T ratio to the power 0.5, while keeping a constant geometry brought out a small but distinct scale dependence in which the transitional flow number varies with tank diameter to the power 0.2. A correlation valid for all geometries and scales is shown in Fig. 48, in which the scale effect has been expressed in dimensionless form by using the square of the Reynolds number divided by the Froude number. The plot shown essentially represents a graph of the average superficial gas velocity against the impeller tip speed at which the transition between clinging cavities (below the line) and large cavities (above the line) occurs.

5.3.   Cavity Configurations with Rushton Turbines

The various regimes of cavity formation, which are shown in Fig. 49, can now be considered. A six blade Rushton turbine operating with an increasing gas flow number develops successively the following cavity configurations:

1. Six pairs of vortex cavities.
2. Six clinging cavities.
3. Three large cavities alternately positioned with respect to three clinging cavities, the 3-3 configuration.
4. Two groups of three large cavities with larger and smaller on successive blades.
5. Three bridging cavities arranged symmetrically and filling the space between alternate blades.

The remarkable stability of the 3-3 configuration leads one to

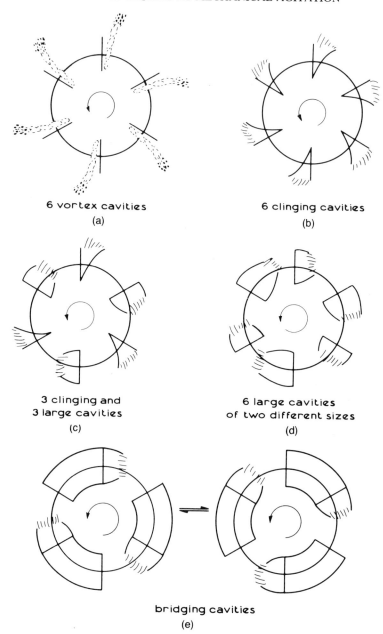

6 vortex cavities
(a)

6 clinging cavities
(b)

3 clinging and
3 large cavities
(c)

6 large cavities
of two different sizes
(d)

bridging cavities
(e)

Figure 49   A six blade Rushton turbine at increasing gas flow numbers: (a) six vortex cavities. (b) six clinging cavities. (c) three clinging and three large cavities; (the 3-3 configuration). (d) six large cavities of alternate sizes. (e) three bridging cavities.

consider the question of possible arrangements with disc turbines having other numbers of blades. Visualization shows that providing there is an even number of blades, then an alternating arrangement is the most stable. Figure 50 shows the configurations adopted by disc turbines carrying eight, four, and two blades. The surprising asymmetric gas loading of the two blade agitator would of course introduce a non-axial loading on the drive shaft which designers might prefer to avoid. It is believed that the transitions in flow configuration of these other impeller designs can be predicted reasonably if the gas flow number is evaluated on the basis of the gas loading per blade instead of the impeller as a whole.

If attempts are made to reproduce similar conditions with disc turbines fitted with an uneven number of blades, no stable cavity arrangements are found. Individual cavities undergo a growth and decline cycle, the frequency of which is determined primarily by the gas rate. All the cavities on a turbine oscillate out of phase with the same frequency with the result that the location at which the largest cavity is found moves around the impeller. For a five blade turbine this location follows a pentagram path.

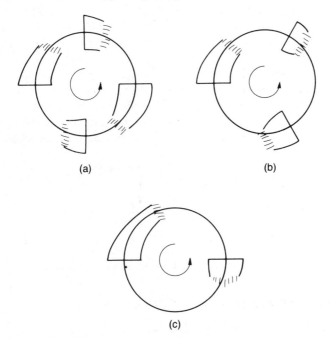

(a)   (b)

(c)

Figure 50   Alternating cavities: (a) an eight blade Rushton turbine. (b) a four blade turbine. (c) a two blade turbine.

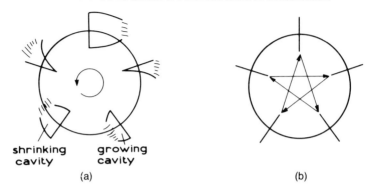

shrinking   growing
cavity   cavity
(a)   (b)

Figure 51   (a) Cavity arrangement on a five blade disc turbine. (b) Changing location
of the largest cavity on a five blade turbine.

Figure 51a shows the cavity arrangement and Fig. 51b shows the shifting location. The frequency of the oscillation in a laboratory rig lay between 0.1 and 1 Hz depending on the gas rate, in any case generally lower than the stirrer speed. Such operator-dependent low frequencies producing nonaxial loading on a shaft must of course be avoided in equipment designed to run at supercritical shaft speeds.

A three blade impeller gives a similar instability, and presumably the unstable nature of the three symmetrical bridging cavities is related to this.

The size of the stable cavities generated under given operating conditions is determined more by the stirrer speed than by the gas rate (Fig. 52). The faster that a blade is moving through the liquid, the larger is the "hole" left behind it. This improves the streamlining and in consequence leads to a lower relative power demand.

(a)   (b)

Figure 52   Stable cavity forms at different stirrer speeds.

5.4.   Hub-mounted Radial Flow Turbines

In contrast to the extensive literature on Rushton turbines, little has been published on the behaviour of flat blade turbines which are mounted directly on the shaft or from a small hub. The six-blade paddle is an intermediate form between the disc turbine and the pitched blade impeller which has been in use since antiquity. Agricola[32] shows (Dover edition, p. 299) how these impellers were used for a three phase mixing operation, extracting metals from ores in suspension in water with mercury.

Examination of cavity formation with flat blade paddles has shown, contrary to all expectation, that even large cavities are six-symmetric in this system, that is, each blade is equally loaded under given conditions and there is no interaction between successive blades that leads to the alternate cavities of the disc turbine. It follows that design based on an uneven number of blades will not experience the same instabilities as were discovered with disc turbines. However the likelihood of gas bypassing the impeller along the shaft has led to disc mounted turbines generally being preferred for dispersion applications.

5.5. Pitched Blade Turbines

5.5.1. Suspension and Gas Dispersion

The choice between a pitched blade agitator and a Rushton turbine is often determined by the requirement of maintaining a solid in suspension. For straightforward gas dispersion applications a Rushton turbine is usually specified, whereas the problems of ensuring that no settling out occurs is usually decisive in leading to the choice of designs intended to produce an axial flow. This is the main reason that extensive work has been published on the suspension characteristics of axial flow turbines whereas there has been little published concerning their gas dispersion performance.

Chapman et al.[13] have pointed out that in many applications the outflow from a pitched blade turbine has a strong radial component; in consequence they prefer the name mixed flow turbine for this class of impeller. However in the present article we will continue to refer to these as axial flow devices. Three phase systems are exploited in very many processing applications. Fermentors and reactors involving suspended catalyst are perhaps the more frequently encountered examples. The combination of the strongly developed roll vortices leaving the impeller tips that were described in Sec. 3.5, together with the large displaced liquid volumes is encouraging for these uses. The two phase flow around these agitators as well as the gassed power consumption and mass transfer properties of these agitators has been recently described by Warmoeskerken et al.[28]

5.5.2. Two Phase Flow Operation

When the agitator is aerated, various hydrodynamic regimes can be distinguished[19]. As is the case with a Rushton turbine, at low gas rates the trailing vortex is filled with small bubbles. However with an axial flow turbine the spinning vortex system is less intensely developed and not all the bubbles passing through the impeller plane are captured by the vortices. There remains a downflow of small bub-

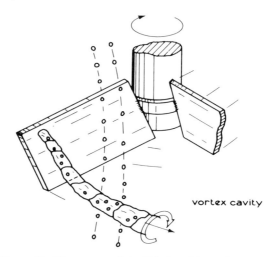

Figure 53   Flow over a pitched blade turbine at low gas rates.

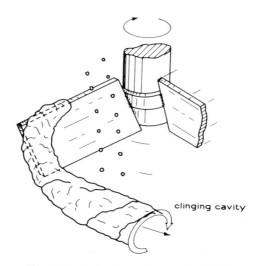

Figure 54   A clinging cavity on a pitched blade.

bles entrained in the considerable convected downflow nearer the turbine axis (Fig. 53). Increasing the gas flow leads to the formation of clinging cavities: the gas-filled vortex attatches completely to the blade (Fig. 54). In this case a downflow of small bubbles can still be detected near the agitator shaft. If the gas rate is increased still further, then large cavities are formed (Fig. 55) though to start with these are in a developing state with some residual vortex motion at the rear where the breakaway and dispersion takes place. Beyond this large cavities with the characteristically flat rear surfaces become stable (Fig. 56). Both the

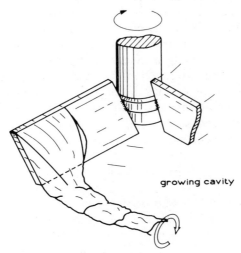

Figure 55   An incipient large cavity on a PBT.

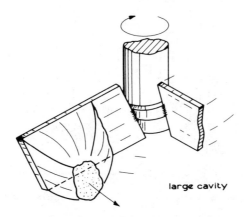

Figure 56   A developed large cavity.

developing and fully formed large cavities are efficient in capturing the gas that passes through the impeller region, and no separate small bubbles can be seen to be convected with the central axial flow.

It has been mentioned above that there are additional geometric effects that are important in pitched blade impeller systems over and above those which affect disc turbines. The most significant of these is the manner of introduction of the gas into the system. The single phase flow of these turbines may include a significant upward component near the tank centre line coupled with the strong outflow form the periphery of the impeller, as may be seen in Fig. 24. Clearly the trajectories of released gas bubbles, particularly small ones, will depend on the exact point of release into the liquid. The change from direct to indirect loading of the impeller, which has been mentioned in passing in Sect. 2.4, has been determined for various arrangements. In general it has been found that the stirrer speed at which the changeover takes place varies with the square root of the volumetric gas rate. Data obtained in a 0.44 m diameter tank for three different six-blade agitators are shown in Fig. 57. The general form of the dependence is

$$N_1(\sin \alpha) = \text{Const.}\ Q_g^{0.5} \tag{4}$$

in which the constant depends on the geometry.

The relevance of the position of the sparger can be seen from Fig. 58. In this case the 0.176 m impeller was mounted at the mid-plane

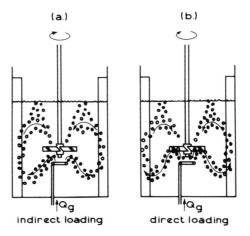

Figure 57   The transition from direct to indirect loading of a pitched blade turbine.

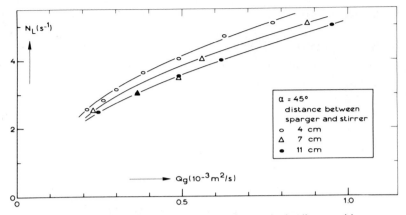

Figure 58   The influence of sparger location on the loading transition.

of the tank. It is no surprise that the greater the separation between the sparger and the impeller, the larger the gas rate that can be introduced before the transition to direct loading takes place. As will be seen this corresponds to maintaining a higher power input level, which for many applications may be an advantage.

5.5.3.   Gassed Power Consumption

The gas loading regime and hence the power demanded by a pitched blade impeller is strongly influenced by the blade angle. In Figs. 59, 60, and 61 the relative gassed power consumption is set out against the gas flow number for various stirrer speeds for impellers with

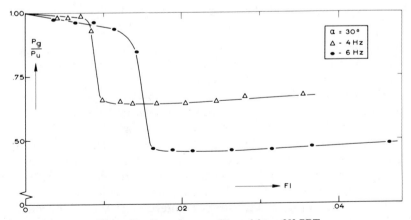

Figure 59   Gassed power demand for a 30° PBT.

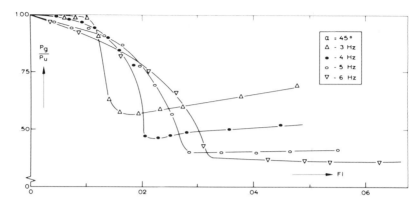

Figure 60   Gassed power demand: 45° PBT.

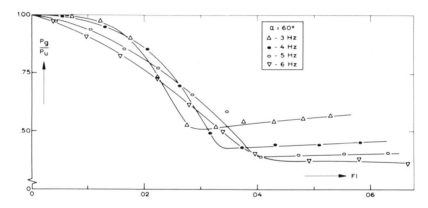

Figure 61   Gassed power demand: 60° PBT.

blade angles of 30°, 45°, and 60°. It seems clear that the sudden drop
in the power demand is coupled to the formation of large cavities,
and in this respect is very similar to the behaviour found with
Rushton turbines. The cavities reduce the pumping capacity of the
impeller. This effect is in addition to the buoyancy forces mentioned
above, so that even if the downflow was originally sufficient to
maintain an indirect loading regime, once the cavities are formed
the system will almost inevitably change to directly loaded operation. The balance of present evidence is that large cavities are first
formed with gas that is recirculated from above the impeller.

   The form of these power curves is interesting. The 60° impeller
shows a less sudden drop in power demand than is found with the
30° turbine, and is in fact much nearer to the behaviour of a

Rushton turbine. However in all cases the sharp drop of the demand is associated with a minimum in the curve; beyond this point there is a steady increase in the input power level. Not only is the point at which the transition occurs affected by the stirrer speed, but the subsequent rise in the demanded power level is also influenced. The result of this is that the power curves for a given turbine intersect twice, in contrast to the case with a Rushton turbine where only a single intersection is found in the power curves. It is not entirely clear how an increasing gas rate can lead to an increase in power demand, but it seems to imply that the pumping capacity of a pitched blade turbine is somehow improved by the presence of larger gas bubbles behind the blades.

6. TURBINE PERFORMANCE AT HIGH GAS RATES

At high gas loadings the pumping action of the agitator ceases to dominate the circulation pattern as a whole, and the effects of buoyancy forces become increasingly important.

6.1. Bridging Cavities

As was pointed out in the section dealing with the alternate cavities that are formed behind the blades of a Rushton turbine, the larger cavities grow rather faster than the smaller ones. The result of the development is that a point is reached when the bigger cavities extend the whole distance from one blade to the other (Fig. 49e). Such bridging cavities restrict the discharge from the turbine to the extent that there is a significant reduction in the pumping capacity. The instability that develops with the three blade stirrer is reflected in the similar hunting of the position of these bridging cavities.

6.2. Flooding Criteria

There are two distinct transitions where the effects of overloading the impeller with gas become evident. These have each been used as the basis for a definition of flooding at one time or another. With an agitator operating at a fixed speed, the first evident change occurs when the circulation in the lower loop becomes insufficient to carry gas with it. The absence of gas recirculated below the impeller plane was considered by Nienow and Wisdom[33] to represent the onset of the flooding transition. These conditions are represented in the changes between Figs. 9e and 9c for a small impeller and between Figs. 9d and 9b for a large impeller, criteria which are clearly different in these two cases. A slightly different view was that of Rushton and Bimbinet[34] who in effect considered the impeller to be flooded even though enough dispersion was taking place as to allow the upper part of the tank to function as a bubble column. This is a somewhat more severe requirement since it excludes the possibility of significant recirculation above the impeller plane, though in normal circumstances that would be difficult to evaluate.

Another criterion which is met in the literature is provided by observation of whether the dispersed gas reaches the vessel wall. This simple test is naturally subject to the criticism that it is too equipment dependent.

Warmoeskerken et al.[35] have put forward the argument that the failure of the pumping action of the impeller provides a good basis for defining the transition, and in this they follow the line taken by Dickey[36]. This transition is represented as that between Figs. 9(b) and 9(a).

Detection of the change in flow direction of the liquid discharged from a turbine is relatively straightforward. A small propeller mounted in the main outflow from the turbine will be driven by the moving liquid so that the changing velocity can be sensed (Fig. 62). In the absence of gas, the rotation of the propeller varies linearly with the speed of the stirrer. Under gassed conditions, no movement is detected by the propeller as long as the stirrer speed is below a

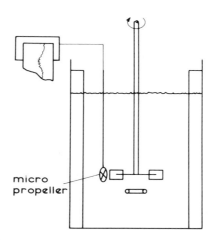

Figure 62   Propeller detector in turbine outflow.

well defined value, beyond which the propeller rotation again increases more or less linearly (Fig. 63). Extrapolation of points to lower stirrer speeds allows accurate determination of the transition.

It is evident that some standardisation of nomenclature is needed. It seems reasonable to describe the operation of an impeller that succeeds in dispersing gas (as represented in Figs. 9c and 9d as "loaded", whilst its inefficient working with an excessive gas supply (as in Fig. 9e), can be called "flooded".

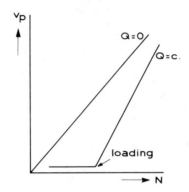

Figure 63   Relationship between propeller rotation rate and stirrer speed, gassed and ungassed conditions.

6.3.   Froude Number and Gas Flow Number at Flooding

The sensitive and accurate detection of the transitions between these two conditions has been used to develop the data of Fig. 64. Here the gas flow number for the changeover has been plotted against the agitator Froude number for a range of experiments carried out in tanks of three quite different sizes. The two regimes are seen to be separated very clearly by a linear relationship between the two variables. The onset of flooding by this criterion is then given by

$$Fl = 1.2 \, Fr \qquad (5)$$

It is interesting that the relationship found by Goossens and Smith[37] for the volume of water raised to the surface of a reservoir by the

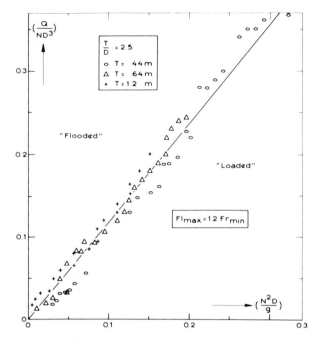

Figure 64   The flooding transition, gas flow number and Froude number.

release of gas at depth is of the form

$$Q_w^3 = 4 \cdot 10^{-4} \, Q_g \cdot gH^5 \tag{6}$$

It seems reasonable to assume that the flooding condition in a stirred tank will be controlled by the ratio between the pumping capacity of the stirrer, given by

$$Q_p = \text{const.} \, N \cdot D^3 \tag{7}$$

and the buoyancy induced flow. With the assumption that the relevant length dimension scales with the tank diameter, this ratio is a linear function of the Froude number according to this reasoning, a fact which is confirmed by the experiment.

The involvement of the Froude number as an appropriate dimensionless group has earlier been suggested by Zwietering[38] and Mikulcova et al.[39] as well as Zlokarnik[40] and Judat[41]. Wiedmann[42] has also used a Froude number to describe the change in pumping regime. The results shown in Fig. 64 support the linearity of the relation between flow number and Froude number, though the work of the various authors would suggest different constants for Eq. (5). The value of 1.2 is specific for a ring sparged flat bottomed tank with the agitator of one third of the tank diameter mounted at its own diameter above the bottom.

An appropriate combination of Froude and Reynolds numbers in a given system may be used to accommodate scale effects. However, the transition predicted here is based on data obtained in tanks with the proportion $T = H$ and is not necessarily valid for other geometries such as tower fermentors with multiple impellers mounted on a single shaft.

6.4. Power Demand at the Flooding Transition

Figure 65 shows two examples of power curves measured around the flooding transition. In this case the presentation favoured by Neinow et al.[43] has been used, with the gas rate being kept constant and the stirrer speed varied. The jump in the power demand occurs at the point of flow transition. This must reflect a sudden increase in the liquid pumping action of the impeller. It is likely that with these conditions the excessively large bubbles that are breaking away upwards through the impeller are no longer blanketing the blades as effectively, and that at least the lower parts of the impeller resume effective displacement of fluid. It should be noted that the gas flow

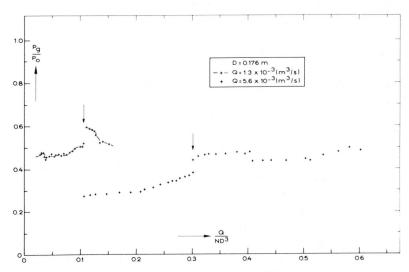

Figure 65 Power curves at flooding point for two different constant gas rates.

rates used for Fig. 65 were not accurately measured and that the values of the constant appropriate to Eq. (5) differ from the 1.2 that has been well substantiated in other extensive experiments.

7. POWER DEMANDS OF ALTERNATIVE GEOMETRIES

An extensive study of the power demands of gassed turbines which differ from the standard Rushton and pitched blade patterns was carried out by van't Riet and Smith[44]. The starting point for the investigation was the knowledge of the trailing vortex system in single phase systems, and information about the various cavity types that are found with disc turbines. It seemed reasonable to expect that if the vortex regime could be extended to higher values of the gas flow number then there would be two benefits. The first of these was better gas phase mixing as a result of the more frequent coalescence and redispersion. A second point was improved mass transfer performance resulting from the smaller mean bubble size produced by the efficient elongational spinning breakaway from the cavity tail. With this in mind, modified disc turbine agitators with many more blades—nine, twelve, and eighteen—were used. Other designs tried included one with perforated blades, aiming to improve dispersion as a result of the impact of the jets that come through the holes in the blades onto the relatively smooth face of the rear of large cavities, and another pattern that was based on blades cut from pipe sections so that the impeller could be driven with either the hollow or convex faces forward. These agitators are shown in Figs. 66, 67, 68,and 69.

The maintenance or otherwise of liquid pumping rate as gas flow increases is reflected in the changes in the power demand relative to that for an ungassed system. The performance of the twelve- and eighteen-blade agitators is shown in Fig. 70. It is immediately clear that these multiblade agitators maintain a far higher power demand on aeration than the standard versions. A similar results is found for the hollow faced turbine. For each of these designs the relative drop in power consumption corresponds to that which would be reached with about three times the gas flow number with a standard agitator.

It is worth pointing out that there are two distinct advantages in working with agitators with these rather flat power characteristics. Firstly, it follows that the mechanical design of the installation can be simplified somewhat, since the motor, gearbox, and shaft loadings change less as the operating conditions are altered. The second

Figure 66   A twelve blade agitator.

Figure 67   The eighteen blade agitator.

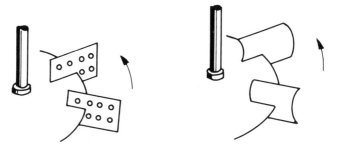

Figure 68   Perforated blade design.   Figure 69   Impeller with hollow/convex blades.

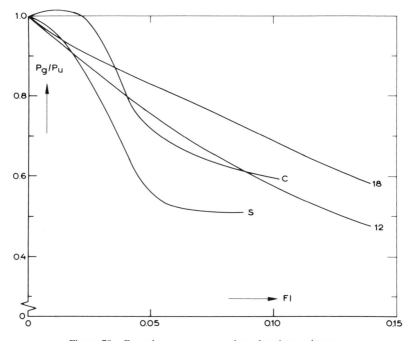

Figure 70   Gassed power consumption of various agitators.

point is that all attempts to improve the mass transfer efficiency by modifying the impeller design lead to the conclusion that in the final analysis only the power input per unit volume is decisive, and therefore any impeller type that maintains a high power input during aeration is attractive.

8. GAS DISPERSION IN APPLICATIONS

Since the principal objective of this article is to present aspects of the hydrodynamics of gas dispersion in stirred tanks only a few examples will be given of situations in which the impeller operation clearly has a direct and immediate effect on the performance of the equipment.

8.1. Gas Holdup

Literature data, for example, that of Calderbank[45] have generally been limited to the water/air system, and attempts are usually made to relate hold-up to the specific power consumption. Information on other systems is also available, for example, Sridhar and Potter[46], who applied a Calderbank type equation to a system involving cyclohexane, though in common with other authors the data correlation has a spread of some 25% (Fig. 71). That work also confirms the reasonable linear logarithmic relationship between holdup and stirrer speed reported by many earlier sources.

Recent results[97] elucidate at least in part the scatter in the earlier data. Figure 72 shows hold-up curves which were determined using an overflow with an external recirculation loop. This technique is sensitive, reproducible, and accurate. There is no doubt that the kink in the curve corresponds to the establishment of the three large

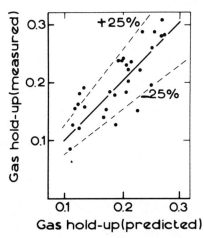

Figure 71   Typical accuracy of generalised correlation for gas holdup, six blade disc turbines (from[46]).

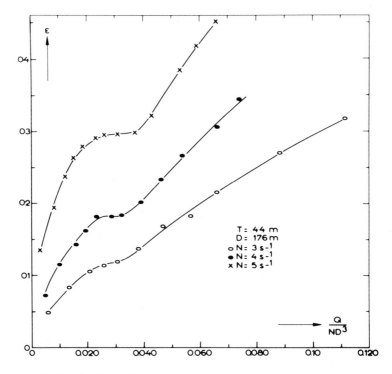

Figure 72   Details of the relation between gas flow number and gas holdup fraction.

cavities behind the blades of a six blade Rushton turbine, and reflects the consequent changes in the dispersion and its distribution.

Adopting different correlating equations for the two hydrodynamic regimes allows the prediction of hold up with much greater accuracy, at least in this particular apparatus, which was for a T of 0.44 m and D of 0.17 m. The equations suggested are:

for Fl less than 0.025 (i.e., no large cavities, Fig. 73.)

$$\epsilon = 0.62 \ Q^{0.45} N^{1.6} \tag{8}$$

and for Fl above 0.025, (Fig. 74)

$$\epsilon = 1.67 \ Q^{0.75} N^{0.7} \tag{9}$$

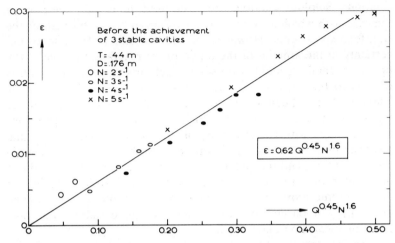

Figure 73   Correlation for holdup, low flow number, Rushton turbines.

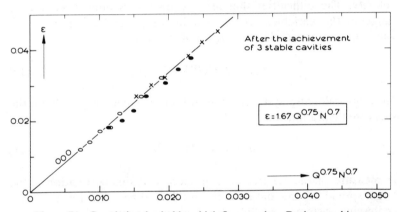

Figure 74   Correlation for holdup, high flow number, Rushton turbines.

8.2.   Mass Transfer

It is usual to consider the mass transfer performance of a stirred tank reactor in terms of the product $k_1 a$. The extensive review by van't Riet[4] considered the techniques that are usually available to determine the mass transfer factor. During the last few years three different methods have been widely used for aqueous systems.

These are (1) physical absorption from air into water which has been purged out with nitrogen, and absorption with catalytically enhanced chemical reaction into either (2) sulphite or (3) hydrazine solutions. Sulphite solutions are often used industrially, because of the ease with which a "standard" solution can be established for the transfer experiments. However the method suffers from some uncertainty in the kinetics of the liquid phase reaction, the role of the necessary catalysts and the possible effects arising from the addition of salts to the absorption system. The hydrozine method, used by Zlokarnik[40] has been shown to give values of the mass transfer factor which are somewhat too high. It appears that the problems are associated with the instability of the copper based catalyst that has been used. It seems certain the these problems will be solved in the near future, and this potentially valuable system, which produces only water and nitrogen as end products, will be widely applied. In the meantime more credence should be placed on results obtained in a nitrogen purged system because of the certainty that only physical absorption plays a role.

It has become usual to correlate mass transfer factors in terms of the power input per unit reactor volume and the superficial gas velocity. The difficulties that arise when this is done over a wide range of the variables are evident in the paper by Smith et al.[48] which quotes correlating equations of the form

$$k_1 a = 0.02(P/V)^{0.475} v_s^{0.4} \qquad (10)$$

for non-coalescing systems while for pure, coalescing systems the suggested equation was

$$k_1 a = 0.01(P/V)^{0.475} v_s^{0.4} \qquad (11)$$

A measure of the relative lack of success of this correlation is the large spread of the data points, no less than 35% for the non-coalescing systems and 20% for the pure systems. For any given conditions the actual data straddle the correlating lines in a curve, and it is impossible to find a single master correlation that can apply to all situations. The analysis is greatly simplified if some account is taken of the hydrodynamic condition of the impeller. By considering mass transfer performance in two distinct regions, depending on whether large cavities are present or not, Warmoeskerken[49] has succeeded in bringing mass transfer data together into two linear relationships of the form, at low gas rates:

$$k_1 a/N = 1.1 \ 10^{-7} \ Fl^{0.6} \ Re^{1.1} \qquad (12)$$

and at high gas rates:

$$k_l a/N = 1.6 \ 10^{-7} \ Fl^{0.42} \ Re^{1.02} \tag{13}$$

Mass transfer has also been studied in tanks agitated with pitched blade impellers. The effectiveness of the mixing can again be described in terms of the relationship between the transfer factor

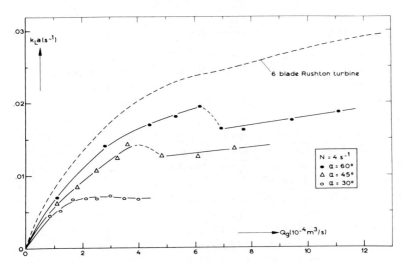

Figure 75   Mass transfer data for pitched blade turbines.

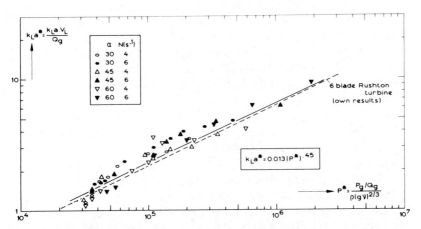

Figure 76   Dimensionless mass transfer against dimensionless power input for pitched blade turbines compared with Rushton turbines.

and the power input. A very similar transition in mass transfer rate is found at the time that the cavity condition changes as is the case with Rushton turbines, as can be seen in Fig. 75. When this performance is defined in terms of the dimensionless group suggested by Zlokarnik, then it can be seen from Fig. 76 that the differences between pitched blade turbines and disc turbines are minimal, and within the normal spread of correlations of this type.

8.3. Suspension with Gas Dispersion

The changes that occur in the liquid flow as a result of the introduction of gas are very evident in the behaviour of suspensions. The chapters concerned with solid–liquid suspensions deal with the criteria that are applied to two phase systems and we should expect that a similar approach could be used for aerated suspensions.

There are many problems that occur in industrial reactors and flotation equipment involving three phases. Frequently there is a loss of performance when gas is introduced, with undesired settling out of solids. It is possible that there are effects on two levels. On the one hand the solids can influence the fundamentals of the dispersion and coalescence processes. As well as this the dispersed gas phase may affect both the large scale circulation and the distribution of turbulence throughout the tank, immediately altering the efficiency with which the solids are maintained off the bottom. Chapman et al.[13] present fundamental information about the performance of agitators operating in three phase systems.

Warmoeskerken et al.[50] have studied the performance of Rushton turbines in these applications, measuring the intensity of light reflected from the bottom of a transparent vessel. The equipment is sketched in Fig. 77. Although the technique has severe limitations, in the particular system studied the last solids to be suspended were always at the centre of the round bottomed tank, and this critical location could be examined by a single point probe.

The detected intensity was related under non-aerated conditions to the normal Zwietering[51] criterion pertaining to off-bottom suspension. Figure 78 shows the results of the experiment. It is found that at the beginning of aeration, the suspending capacity of the agitator is actually increased somewhat, and that only after the formation of the large cavities does the solid settle faster than in the unaerated system. This result is however specific to disc turbines, neither hub-mounted paddles or pitched blade turbines show similar behaviour.

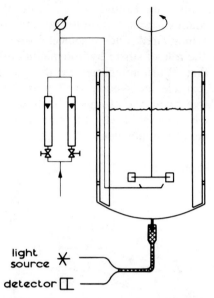

Figure 77   Fibre optic detector for sedimented solids.

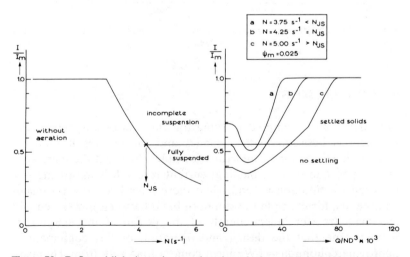

Figure 78   Reflected light intensity as a function of stirrer speed, showing increased suspension of solids with moderate gas flow rates in dished end vessel agitated with a Rushton turbine.

Recent investigations by Frijlink et al.[52] into the relationship between the minimum stirrer speed (and associated power demand) needed to achieve an aerated solids suspension with a pitched blade turbine reinforce the role of stirrer hydrodynamics in this. Figure 79 shows the way that P_{js} is influenced by the gas rate for three combinations of particle and geometry. On the same figure the points at which large cavities are formed are shown. These gas flow numbers are closely associated with the transition between direct and indirect loading of the impeller.

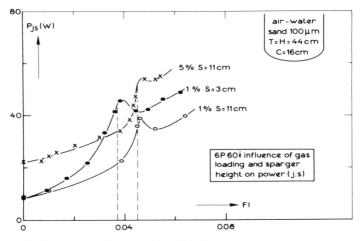

Figure 79   Power demand to maintain solids in just suspended condition in aerated vessels.

9. CONCLUSIONS

This article has attempted to bring together various aspects of the interactions between hydrodynamics and process result that are relevant in stirred tank contactors. It will be clear that there are very many open questions remaining; current research is improving our knowledge continuously, and while there have been several disappointments, for example the stubborn bond between power demand and mass transfer independently of the design of the agitator, it remains true that the design and operation of this equipment is improving continuously. We have come a long way from Thomson who in 1855 thought that the £43.14s.3d. remaining from his original grant from the Royal Society would be enough to finance "the more extended experiments yet remaining to be made" which would be "for the sake of greater steadiness of motion, worked by steam power instead of the hand of an operator."

10. NOTATION

A centripetal acceleration, ms^{-2}
A_t centripetal acceleration at blade tip, i.e., $2(\pi N)^2 D$, ms^{-2}
a specific contact area per unit volume, m$^2 \cdot$m^{-3}
D agitator diameter, m
g gravitational acceleration, ms^{-2}
H liquid depth, m
k_1 mass transfer coefficient, m^{2}s^{-1}
N stirrer speed, s^{-1}
N_{js} stirrer speed at which solid is just suspended, s^{-1}
N_1 stirrer speed of loading–flooding transition, s^{-1}
P power input to mixer shaft, W
P_g power input, gassed system, W
P_{js} power input with solids just suspended, W
P_u power input, ungassed system, W
Q_g gas input flow rate, m^{2}s^{-1}
Q_p liquid pumping rate of agitator, m^{3}s^{-1}
Q_p liquid pumping rate of agitator, m^{3}s^{-1}
Q_w water raised to pool surface by bubble column, m^{3}s^{-1}
T tank diameter, m
V liquid volume in stirred tank, m^3
v_t blade tip velocity, ms^{-1}

Greek

ϵ volumetric gas fraction, ms^{-1}
μ liquid phase viscosity, Pa s
ρ liquid phase density, kg m^{-3}
Fl gas flow number, Q_g/ND^3
Fr Froude number, $N^2 D/g$
Po Power number, $P/N^3 D^5\rho$
Re Reynolds number, $ND^2\rho/\mu$.

REFERENCES

1. Midoux, N., and Charpentier, J. C. (1979). Les reacteurs gaz–liquide, cuve agitee mecaniquement, *Entropie* **88**, 5–38.

2. Brauer, H. (1979). Power consumption in aerated stirred tank reactor systems, *Adv. Biochem. Eng.*, **13**, 87–117.
3. Schugerl, K. (1982). New bioreactors for aerobic processes. *Int. Chem. Eng.*, **22**, 591–610.
4. van't Riet, K. (1979). Review of measuring methods and results on non-viscous mass transfer in stirred vessels, *Ind. Eng. Chem. Proc. Des. Dev.*, **18**, 357–364.
5. Charpentier, J. C. (1981). What's new in absorption with chemical reaction, *Trans. Inst. Chem. Eng.*, **60**, 131–156.
6. Hinze, J. O. (1955). Fundamentals of the hydrodynamic mechanism of splitting in dispersion processes, *AIChE J.* **1**, 289–295.
7. Cutter, L. A., (1966). Flow and turbulence in a stirred tank, *AIChE J.*, **12**, 35–45.
8. Nagata, S. (1975). *Mixing, Principles and Applications*, Kodansha, Wiley, New York.
9. Rushton, J. H., Costich, E.W., and Everett, H. J. (1950). Power characteristics of mixing impellers, *Chem. Eng. Prog.*, **46**, 395–476 and 467–476.
10. Kipke, K. D. (1982). Fluid loadings on impellers and tanks in gassed liquids, *Proc. 4th European Mixing Conf.*, Nordwijkerhout, The Netherlands, 355–370. BHRA Fluid Engineering, Cranfield, Bedford, England.
11. Thomson, J. (1855). Experiments on the friction of discs revolving in water, *Proc. Roy. Soc.*, **7**, 509–511.
12. Dickey, D. S., and Fenic, J. G. (1976). Dimensional analysis for fluid agitation systems, *Chem. Eng.*, Jan. 5, 139–143.
13. Chapman, C. M., Nienow, A. W., Cooke, M., and Middleton, J. C. (1983). Particle–gas–liquid mixing in stirred vessels, *Chem. Eng. Res. Des.*, **61**, 71–81; *ibid.* 82–95.
14. Chapman, C. M., Nienow, A. W., and Middleton, J. C. (1981). Particle suspension in a gas sparged Rushton turbine agitated vessel, *Trans. Inst. Chem. Eng.*, **59**, 134–137.
14. Takeda, K., and Hoshino, T. (1966). Flow patterns of liquid inside the turbulent mixing impellers, *Kagaku Kogaku*, **4**, 394–398.
15. Beckner, J. L. (1966). Mixing with anchor agitators, Ph.D. Thesis, University of Wales, Cited by Peters and Smith[16].
16. Peters, D. C., and Smith, J. M. (1968). Fluid flow in the region of anchor agitator blades, *Trans. Inst. Chem. Eng.*, **45**, T360–366.
17. Smith, J. M., (1972). Alternative flow regimes in agitated vessels, *The Chem. Engr. No.*, **261**, 182–185.
18. van't Riet, K. (1975). Turbine agitator hydrodynamics and dispersion performance, Doctoral Thesis, T.H. Delft.
19. Warmoeskerken, M. M. C. G., Spew, J., and Smith, J. M. (1984). Gas–liquid dispersion with pitched blade turbines, *Chem. Eng. Commun.*, **25**, 11–29.
20. Metzner, A. B., and Otto, R. E. (1957). Agitation of non-Newtonian fluids, *AIChE J.*, **3**, 3–10.
21. Gunkel, A., and Weber, M. E. (1975). Flow phenomena in stirred tanks, *AIChE J.*, **21**, 931–949.
22. van't Riet, K., Bruijn, W., and Smith, J. M. (1976). Real and pseudo turbulence in the discharge stream from a Rushton turbine, *Chem. Eng. Sci.*, **31**, 407–412.
23. Popiolek, Z., Whitelaw, J. H., and Yianneskis, M. (1984). Unsteady flow over disc turbine blades, *Proc. 2nd Internat. Symp. Applications Laser–Doppler Anenometry to Fluid Mechanics*, Lisbon, Portugal. Paper 17-1.
24. Tatterson, G. B., Hsien-hwa, S. Y., and Brodkey, R. S. (1980). Stereoscopic visualization of the flows for pitched blade turbines, *Chem. Eng. Sci.*, **35**, 1369–1375.
25. Fort, I., Medek, J., Placek, J., and Hulanek, M. (1975). Hydraulic characteristics of paddle impellers with flat inclined blades, *Coll. Czech. Commun.*, **40**, 3443–3458.

26. van't Riet, K., and Smith, J. M. (1973). The behaviour of gas–liquid mixtures near Rushton turbine blades, *Chem. Eng. Sci.*, **28**, 1031–1037.
27. Matsamura, M., Masunaga, H., and Kobayashi, J. (1977). A correlation for the flow rate of gas entrained from the free surface of an aerated tank, *J. Ferm. Technol.*, **55**, 388–400.
28. Warmoeskerken, M. M. C. G., and Smith, J. M. (1982). Description of the power curves of turbine stirred gas dispersions, *Proc. 4th European Conf. on Mixing*, Nordwijkerhout, The Netherlands, 237–246. BHRA Fluid Engineering, Cranfield, Bedford, England.
29. Nienow, A. W., Chapman, C. M., and Middleton, J. C. (1977). Gas recirculation rate through impeller cavities and surface aeration in sparged agitated vessels, *Chem. Eng. J.*, **17**, 111–118.
30. Bruijn, W., van't Riet, K., and Smith, J. M. (1974). Power consumption with aerated Rushton turbines, *Trans. Inst. Chem. Eng.*, **52**, 99–104.
32. Agricola, G. (1956). *De Re Metallica*, trans. Hoover and Hoover, Dover 1950.
33. Nienow, A. W., and Wisdom, D. J., (1976). Mixing in two phase systems, *I. Chem. Eng. 3rd Annual Research Meeting.*
34. Rushton, J. H., and Bimbinet, J. J. (1968). Holdup and flooding in air–liquid mixing, *Can. J. Chem. Eng.*, **46**, 16–21.
35. Warmoeskerken, M. M. C. G., and Smith, J. M. (1984). The flooding transition with gassed Rushton turbines, Fluid Mixing II, *Institution of Chemical Engineers Symposium Series* **89**, 59–67.
36. Dickey, D. S. (1979). Turbine agitated gas dispersion—power, flooding and holdup. *AIChE 72nd. Annual Meeting*, San Francisco, Paper 116d.
37. Goossens, L., and Smith, J. M. (1975). The hydrodynamics of unconstrained bubble columns for mixing lakes and reservoirs. *Chem. Ing. Tech.*, **47**, 951, MS 301/75.
38. Zwietering, T. N. (1963). Contribution to discussion after Design of Agitators for Gas–liquid Contacting, K. R. Westerterp and H. Kramers De Ingenieur, Chem. Tech. 6, 18/10/63, ch 55–ch 59 reported on ch 60–ch 61.
39. Mikulcova, E., Kudrna, V., and Vlcek, J. (1967). Stanoveni Meznich Podminek Zahlceni Michadla (Conditions for Impeller Flooding), Scientific Papers, Inst. of Chemical Technology, Prague, **KI**, 167–183.
40. Zlokarnik, M. (1978). Sorption characteristics for gas–liquid contacting in mixing vessels, *Adv. Biochem. Eng.*, **8**, 133–151.
41. Judat, H. (1979). Begaasen von niedrigviskosen Flussigkeiten *Chem. Ing. Tech.*, **51**, 710–716.
42. Wiedmann, J. A. (1983). Zum überflutungsverhalten zwei- und dreiphasig betriebener Rürreactoren, *Chem. Ing. Tech.*, **55**, 689–700.
43. Nienow, A. W., Wisdom, D. J., and Middleton, J. C. (1977). The effect of scale and geometry on flooding, recirculation and power in gassed stirred tanks, *Proc. 2nd European Conference on Mixing*, Cambridge, Bedford, England, Fl 1–16. BHRA Fluid Engineering, Cranfield, Bedford, England.
44. van't Riet, K., and Smith, J. M. (1974). Fluid flow and gas dispersion in turbine agitated tanks, *Dechema Monographien*, **74**, 93–101.
45. Calderbank, P. H. (1958). Physical rate processes in industrial fermentation, part I: The interfacial area in gas–liquid contacting with mechanical agitation, *Trans. Inst. Chem. Eng.*, **36**, 443–463.
46. Sridhar, T. and Potter, O.E. (1980). Gas holdup and bubble diameters in pressurised gas–liquid stirred vessels, *Ind. Eng. Chem. Fundam.*, **19**, 21–26.
47. Warmoeskerken, M. M. C. G., Feijen, J., and Smith, J. M. (1981). Hydrodynamics and power consumption in stirred gas–liquid dispersions, *I. Chem. Eng. Symp. Ser. No. 64.*, Fluid Mixing, J1–J14.
48. Smith, J. M., van't Riet, K., and Middleton, J. C. (1977). Scale up of agitated gas–liquid reactors for mass transfer, *Proc. 2nd. European Conference on Mixing*,

Chapter 6

Gas–Liquid Mixing and Mass Transfer in High Viscosity Liquids

ALVIN W. NIENOW

*Department of Chemical Engineering, University of Birmingham,
Birmingham B15 2TT, UK*

and

JAROMIR J. ULBRECHT†

*Department of Chemical Engineering, State University of New York, Buffalo, NY 14260,
USA*

1. INTRODUCTION

In a number of technologies, gas is dispersed in very viscous and often rheologically-complex liquids. Examples of this requirement are found in the polymer and fermentation industries. In both of these examples, the process is generally conducted batchwise. Initially the fluid is often Newtonian with a viscosity close to that of water and becomes increasingly viscous, pseudoplastic and viscoelastic with the passage of time.

The change of rheological properties affects gas and liquid flow patterns, gas dispersion properties, the mechanism of bubble break-up, the relationship between agitation conditions and gas–liquid mass transfer rates and the power drawn by the agitator for any particular combination of rotational speed and gassing rate. All of

†Present address: Center for Chemical Engineering, National Bureau of Standards,
Washington D.C. 20234, USA

these phenomena interact with each other and ideally such inter-actions should be taken into account.[1]

Most of the fluids of industrial importance are shear-thinning and rheological models for this type of behaviour are discussed in detail in Chapter 4. For completeness, three are mentioned here. Shear stress-shear rate data may be correlated by a Herschel–Bulkley type model, that is,

$$\tau = (\tau_y)_{HB} + K_{HB}\,\dot{\gamma}^{n_{HB}} \tag{1}$$

which degenerates to a power law model if the fluid does not exhibit a yield stress, that is,

$$\tau = K\dot{\gamma}^n \tag{2}$$

where both n and n_{HB} are less than 1. These models are discussed in Chapter 4. Another model, the Casson model, has been frequently used for describing suspensions of filamentous mycelia as found in a number of fermentations[2],

$$\tau^{1/2} = (\tau_y)_C^{1/2} + K_C\dot{\gamma}^{1/2} \tag{3}$$

The Casson model does not, however, meet the requirements of material invariance and, therefore, its use for the interpretation of the spatial state of stress is ambiguous.

The yield stress is difficult to determine unequivocally and yet, as shown below and in Chapter 4, it may be of considerable importance. From graphs based on the Casson equation, it is very easy to get an estimate of a yield stress, $(\tau_y)_C$.

The fluids may also exhibit viscoelasticity. Again this phenomenon is discussed in detail in Chapter 4. For the purposes of this chapter, the following brief resumé is a useful reminder of the properties of most importance. Firstly, these fluids exhibit forces normal to the plane of shear. Experimental data relating to these forces may be correlated conveniently by a power law relationship between the first normal stress difference, ν_1, and the shear rate, $\dot{\gamma}$, by

$$\nu_1 = A\dot{\gamma}^b \tag{4}$$

Secondly, such fluids are sometimes called memory fluids in that their behaviour at any instant may depend on their previous experience. In particular, their behaviour will depend on the relaxation time of the fluid, λ, as compared to the speed and frequency of the deformation it is undergoing. Thirdly, the extensional viscosity is dramatically increased as compared to inelastic fluids and may be many hundreds of times greater than the shear viscosity.

In such high viscosity fluids, very tiny gas bubbles are constrained to move with the fluid (including not moving at all, if the fluid is stationary at any point which is particularly probable if it exhibits a yield stress). The very presence of these tiny bubbles means that, in many cases, it is the rheological properties of this stable or equilibrium two phase mixture which are most important. The rheological properties of the combined phases have, however, not been taken into account in any of the published work to date.

Chapters 4, 5, and 8 show the great value of flow visualisation as an aid to understanding the complex flows that can occur when mixing rheologically-complex fluids and two phase systems. The same benefit also accrues from this technique when studying gas dispersion in high viscosity fluids. A particularly useful group of fluids which covers a wide range of rheological properties in a coherent fashion are as follows: corn syrup (Newtonian); Boger fluid[3] (corn syrup plus traces of polyacrylamide) has a shear independent viscosity but it is strongly viscoelastic; Carbopol which is very shear-thinning and has a large yield stress, but is only mildly viscoelastic or not at all depending on the pH; and sodium carboxy methyl cellulose (CMC) of which there are many types, but those with high molecular weights give solutions which are strongly shear-thinning without a yield stress and viscoelastic.

Table 1 shows some typical rheological properties of two groups of solutions of the above fluids as measured on a Weissenberg rheogoniometer. Also included are the rheological properties of two Xanthan gum solutions. Each group of fluids gives rise to similar Reynolds numbers (estimated by the technique of Metzner and Otto[4] as discussed in Chapter 4 and below) under the conditions typically found in agitated vessels. It can be seen that the rheological properties of Xanthan solutions are reasonably matched by one or other of the transparent model fluids.

Xanthan is produced by fermentation and is an important product with many uses[29]. The rheological properties of the fermentation broth are almost entirely dependent on the high molecular weight gum which is exuded into it. Similarly, it is the high molecular weight polymers which give rise to the rheological properties typically found in the polymer-based industries. Thus, these transparent model fluids are very suitable for studying gas–liquid mixing for these industrially-important opaque fluids.

However, in many fermentation processes, the measured rheolo-

Table 1   Rheological Parameters of Fluids (25°C)

	Fluid and concentration (% w/w)	Shear stress parameters		1st Normal stress diff. parameters		Yield stress	
		K	n	A	b	$(\tau_y)_{HB}$	$(\tau_y)_C$
		$(N s^n m^{-2})$		$(N s^b m^{-2})$		$(N m^{-2})$	$(N m^{-2})$
High viscosity	4.5% Xanthan	50.0	0.13	6.73	0.46	8.3	19.4
	1.4% CMC	16.6	0.41	6.37	0.52	0	0
	0.30% Carbopol (pH 4.4)	31.9	0.24	1.54	0.61	19.1	15.0
	0.05 wt% PAA in 91% wt corn syrup	3.9	0.95	4.6	1.36	0	0
Low viscosity	2.0% Xanthan	20.0	0.16	2.13	0.57	5.8	11.7
	0.8% CMC	4.11	0.53	2.33	0.57	0	0
	0.17% Carbopol (pH 4.6)	11.2	0.29	0.218	0.84	4.9	7.4
	0.1% wt% PAA in 78% wt corn syrup	1.06	0.91	0.17	1.68	0	0

gical properties of the opaque fermentation broth are almost entirely due to the concentration and morphology of the dispersed solid phase. When such broths are highly-viscous and rheologically-complex, the dispersed phase is usually a suspension of filamentous mycelia. It is by no means certain that, in this case, the model solutions are a fair representation of the real ones of interest. The same difficulty exists when handling gas dispersion under the high solid loadings found in some froth flotation cells. No satisfactory way around this problem has been found as yet.

2.   IMPELLER SELECTION

In Chapter 4, the agitators used in many non-Newtonian technologies are classified into three classes. Of these three, two classes involve relatively slow moving devices which cause fluid motion throughout the viscous fluid either by their positive pumping action, for example, helical screws and helical ribbons; or, by their size and shape which causes them to pass during rotation through the very extremities of the vessel, for example, gates and anchors. The third type operate at high speeds and rely on good momentum transport for mixing. It is clear that for the mixing of very viscous, shear-thinning fluids, relying on good momentum transport may be wishful thinking. This is true, particularly if the fluid has a yield stress: the impeller may just cut out a cavern beyond which the fluid is totally stagnant.

However, as Chapter 5 shows, this type of high speed agitator is just the sort that is used for gas dispersion in low viscosity fluids. Indeed, a certain minimum speed is required if the agitator is to be able to disperse the gas. Thus, the process of dispersing gases into high viscosity fluids sets an impeller two inherently conflicting requirements.

The Rushton disc turbine has certain advantages, especially for batch processes in which the viscosity increases considerably with time. As can be seen from Fig. 5 of Chapter 6, increases of viscosity (or apparent viscosity) of as much as four orders of magnitude in unaerated fluids may not increase at all the power demand of such agitators. It is also good for gas dispersion in low viscosity fluids. On the other hand, it is a low flow, high-shear impeller[5]. For this reason, together with the tendency for disc turbines to form caverns which localise fluid motion to the proximity of the agitator, combining the Rushton turbine with a high flow, low-shear impeller, for example, a 45°-pitched blade agitator, should offer advantages. It is clearly desirable to have the disc turbine as the lower of the two in order to sparge the gas into it. This combination of impeller types may be thought of as a compromise between those impellers which cause mixing by their size and shape and those that do so due to their good momentum transport.

The Intermig agitator (Fig. 1) offers a similar compromise, that is, it is of large diameter and two or more are used on the same shaft. The fact that each agitator only has two blades, each rather narrow in cross section, and that its staggered mini-blade ends are specifically designed on hydrodynamic grounds to minimise local form drag losses, means that it has a lower power number. Thus, the set can be rotated at high speeds without drawing unacceptably high power levels.

Finally, some progress has been made in using two independently driven agitators[6]. One agitator is of the positive pumping type which rotated slowly in a draught tube and moves the fluid around the vessel. The other is rotated at high speed to disperse the gas. The device is shown in Fig. 2. Though the idea is based on sound principles and the technique has been used for filamentous fermentations, more data is needed on it before it can be considered to have proven advantages.

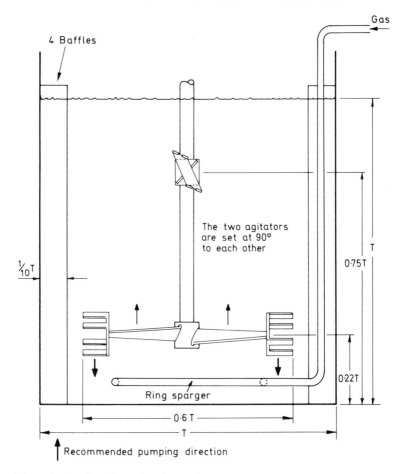

Figure 1   A pair of Intermig agitators for gas dispersion (approximately to scale).

3.   REYNOLDS NUMBER

The method of Metzner and Otto[4] is generally used to calculate Reynolds numbers for high speed agitators in the laminar region. The method is discussed in detail in Chapter 4 and is the one used in this chapter. Thus the Reynolds number, Re, is $\rho ND^2/\mu_a$ where μ_a is the apparent viscosity. For power law fluids, this definition means

$$\text{Re} = \frac{\rho N^{2-n} D^2}{K k_s^{n-1}} \tag{5}$$

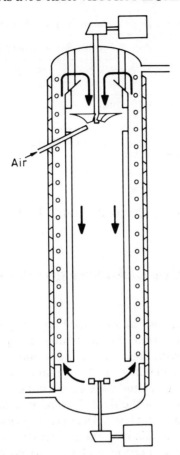

Air

Figure 2 A fermenter with independently driven agitators, one for bulk flow and one for high shear, developed for high viscosity gas–liquid systems[6].

Though of unproven validity in the high Re range the same method is used in this chapter for calculating all Reynolds numbers for shear thinning fluids. The bulk of this chapter refers to fluids in which the Reynolds numbers (defined in terms of the apparent or true dynamic viscosity) are less than about 1000.

Interestingly, low flow behaviour indices change the effect of scale on the Reynolds number compared to Newtonian systems. For Newtonian systems, especially at low viscosities, much higher Reynolds numbers are obtained on scale-up with any of the commonly recommended rules. However, similar Reynolds number are found on both the small and large scale when highly viscous power law

Table 2   Effect of Power Law Behaviour on Impeller Reynolds Number[7]

Scale	Impeller Reynolds number		
	$K = 50^c$ $n = 0.13$	$K = 50$ $n = 1$	$K = 10^{-3\ d}$ $n = 1$
$D = 0.174$ m, $N = 16.75$ s$^{-1\ a}$	980	10	5×10^5
$D = 1.2$ m, $N = 2$ s$^{-1\ b}$	879	58	2.9×10^6

[a] Equivalent to $D/T = 0.6$ impeller used in small scale experiments[16].
[b] Typical industrial scale and speed.
[c] 4.5% Xanthan.
[d] Water.

solutions are studied. This is illustrated by the examples given in Table 2[7].

4. PHILOSOPHY OF GAS DISPERSION IN HIGH VISCOSITY SYSTEMS

In low viscosity fluids, as explained in Chapter 5, break-up of sparged gas occurs from the ventilated cavities behind impeller blades. Once break-up has occurred, gas bubbles move away from the agitator and, depending on the agitator speed, they may recirculate around the vessel and back to the impeller, or they may pass relatively directly out of the fluid. During their passage through the fluid, the bubbles may coalesce. The extent of coalescence will affect the extent of gas recirculation. The important point is that there are two aspects of gas dispersion that need to be considered. Firstly, there is gas break-up in the region of the impeller. Secondly, there is gas–liquid dispersion throughout the remainder of the vessel.

A low viscosity fluid is easily made to move throughout the vessel provided that the impeller is not flooded, but the motion of gas bubbles need not follow that of the fluid. For high viscosity shear thinning fluids (as indicated in Chapter 4), the liquid may not be circulating throughout the vessel even without gas sparging. On the other hand, because the viscosity is high, the gas bubbles are much more likely to follow the liquid. Thus, a somewhat different approach is required as compared to that adopted in Chapter 5 for low-viscosity gas dispersion.

As already suggested in the section in this chapter on impeller selection, dual impellers are commonly used to ensure good mixing. However, to simplify matters, the case of a single agitator will be discussed first in relation to gas-filled cavity formation. As shown in Chapter 5, the gas filled cavity structure determines power demands to a considerable extent and single impeller power consumption is considered next. Finally impeller combinations will be considered with special emphasis on gas dispersion throughout the whole of the vessel.

5. GAS-FILLED CAVITY STRUCTURE

Cavity structure has been studied for all of the transparent fluid indicated in Table 1, plus corn syrup. The observational technique is essentially the same as that indicated in Chapter 5, but it is based on the use of a derotational prism (see Fig. 3)[8] rather than a rotating television camera.

Under unaerated conditions, a vortex structure similar qualitatively to that described in Chapter 5 can be clearly seen for Reynolds numbers greater than about 10 for the fluids indicated in Table 1 [9]. Thus, vortices continue to exist down to much lower Reynolds numbers than the lowest covered by the work of van't Riet[10]. When aerated, gas is attracted into these low pressure regions, just as it is with low viscosity fluids. However, the cavity shape and its stability depend very much on whether the fluid is viscoelastic and on the agitator Froude and Reynolds number.

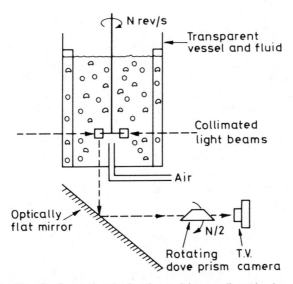

Figure 3   Use of a de-rotational prism for studying gas dispersion by agitators[8].

Considering agitation by disc turbines at low speeds, there is a minimum agitator speed before cavities form in low viscosity fluids. Van't Riet[10] has suggested that a Froude number greater than 0.10 is required before cavities form at all. Nienow et al.[11] have shown that above this speed, "ragged" cavities form similar in shape in plain view to clinging cavities (see Chapter 5), but mainly on the top half of the blades. For water, above a speed given by the equation,

$$Fl_G = 30(D/T)^{3.5} Fr,\tag{6}$$

these ragged cavities generally change to the stable 3-3 cavity structure described in Chapter 5[12].

For Reynolds numbers greater than about 1000, as the viscosity increases so the agitator speed to disperse gas also slowly increases[13]. Typically, a 10-fold increase on viscosity increases N_{CD} by about 50%, that is,

$$N_{CD} \propto \mu^{0.15}\tag{7}$$

However, in other respects the cavity structures that form at different speeds are very similar to those described in Chapter 5.

If the fluid is sufficiently viscous to give Reynolds numbers less than about 10, it appears that high speed agitators are unable to disperse the gas[9, 13].

For Reynolds numbers greater than 10 but less than about 1000, there is an extremely complex interaction involving Reynolds number, agitator Froude number and, to a less extent, gassing rate. As the viscosity increases, a greater speed is required for cavities to be formed as indicated by Eq. (7). However, perhaps because this increasing speed is required to enable cavities to form, the effect of gassing rate progressively decreases as compared to that implied by Eq. (6).

If the fluid is Newtonian or shear-thinning, but inelastic, cavities form with increases in agitator speed above the minimum given by the Froude number criterion in a rather similar way to those found with low viscosity fluids (see Chapter 5 and below). However, only at very very low gassing rates ($\ll 1/4$ vvm)* do vortex or clinging cavities form. At all realistic gassing rates, two groups of large cavities form with larger and smaller ones on successive blades[14].

---

* The "vvm" is a unit of the specific gassing rate (volumetric gas rate per minute per volume of liquid in the tank) often used in fermentation processes.

In addition, changes in gassing rate have little effect on the actual size of these cavities though, if the gas is turned off, they slowly collapse passing through the clinging and vortex cavity form on the way.

If the fluid is inelastic but has a yield stress, the behaviour at the same conditions of aeration and agitation is almost identical to that of other inelastic shear-thinning fluids, but with one difference. Once gas-filled cavities have formed behind the blades, some gas remains entrapped there even after the agitator is stopped. The actual volume of the "permanent" cavities depends on the magnitude of the yield stress. This behaviour can be seen with Carbopol and, by inference, occurs with Xanthan[13].

If the fluids are viscoelastic, for example, the Boger fluids or CMC solutions of Table 1, then a rather different behaviour is observed. A similar minimum speed is again required to enable cavities to form. However, once formed, their shape is different and they have been called split[15] or anvil[16] cavities. A perspective view is indicated in Fig. 4. From below, the cavity appears split and from the side like an anvil. Hence the two names reflect the manner in which the cavities were first observed by the two research groups. In addition to the different shape as compared to inelastic fluids, the cavity is the same behind each blade. Also, the size of each is greater and independent of the gassing rate. Indeed, if the gas is turned off, the cavities continue to remain strapped to each blade with only a slight loss of gas. If the agitator is stopped, the gas disperses.

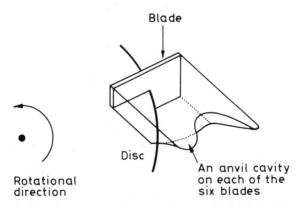

Figure 4   Diagrammatic representation of typical gas-filled cavities on Rushton turbines in high viscosity, viscoelastic liquids[16].

For gas dispersion with the other types of high speed agitators, that is, the angled-blade and the Intermig type, a qualitative picture of cavity formation can be gained from Chapter 5 and the description above. For example, in the case of angled-blade agitators, only at very low gas rates are streamer[17] (or vortex) cavities (Fig. 5a) formed. At all realistic rates, large equal sized cavities are formed on all blades[17] (Fig. 5b). If the fluid is elastic, these large cavities are considerably bigger than when it is not. In addition, they form at all gassing rates and remain buckled to the back of the blade even after the gas is turned off.

The effect of elasticity is even more dramatic with the two-bladed Intermig agitators. In this case, the gas-filled cavities can extend behind the outer twin split blades around a circle of diameter equal

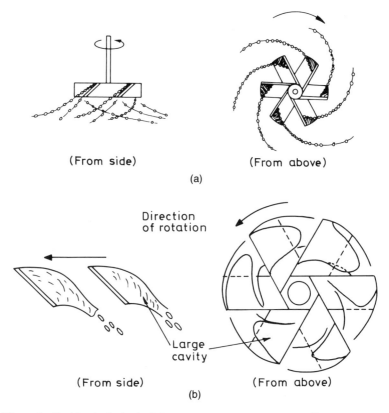

(From side)   (From above)

(a)

Direction
of rotation

Large
cavity

(From side)   (From above)

(b)

Figure 5   Cavities in inelastic fluids on an angled blade agitator[17]. (a) Vortex or streamer cavities at $Q_G \to 0$; (b) Large equal sized cavities, formed at all realistic gassing rates (both downward pumping).

to that of the agitator, almost all the way back to the following blade[14].

It seems most probable that it is the extremely enhanced elongational viscosity, found in viscoelastic fluids, that leads the formation of different gas-filled cavities as compared to inelastic ones. This high elongational viscosity leads to a much more stable cavity structure, reducing the ease with which gas breaks away. This is true even to the point that cavities remain present without additional gas addition. It also causes the cavities to be bigger and, presumably because of the extra forces involved, they all become equal.

6. POWER DRAWN UNDER CONDITIONS OF SPARGING

Since gas-filled cavities do not form for Reynolds numbers less than about 10, the gassed power is equal to the ungassed[13].

For Reynolds numbers between about 10 and 1000, because even at low gassing rates the large cavity structures form very readily, the effect of different gas rates on gassed power number is much less than in low viscosity fluids. This relative insensitivity to gassing rates has now been reported on a number of occasions[13, 15, 18]. Figure 6 shows some typical data for a viscoelastic CMC solution for Rushton turbines.

Interestingly, in some very recent work, Greaves and Loh[19] have found a similar insensitivity when dispersing gas into high-concentration solid suspensions. For low concentration suspensions (30% by wt.), they found relationships between agitator power, speed and aeration rate are very similar to those described in Chapters 5 and 8. However, for slurries of 50% by weight solids and higher, the power drawn was independent of gassing rate. Visual observation was not possible, but it can be inferred that a change in cavity structure took place due to the presence of solids in high concentration.

In contrast to Chapter 5, by reason of this insensitivity to gassing rate, it is more meaningful to link changes in cavity structure with firstly, changes in speed. There interactions can then be linked to changes in gassed power number, Po_g. With Rushton turbines, the following sequence of events occurs; once the Froude number is greater than about 0.1, cavities begin to form. In the first instance, these are like ragged cavities and only grow on the upper edge of the blades. As a result, they are rather small and the power number is only a little less than in the ungassed case. Further increases in impeller speed produce the stable cavity structure described in the previous section, depending on whether the fluid is elastic or not.

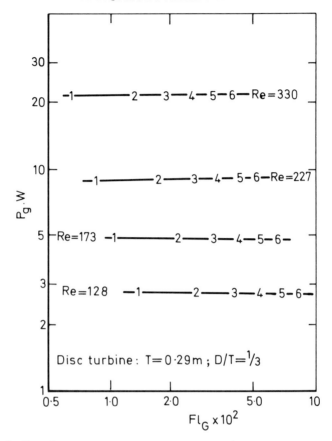

Figure 6   Gassed power data for viscoelastic C.M.C. solution[13]. (1, Q_G = 0.26 vvm; 2, Q_G = 0.54 vvm; 3, Q_G = 0.79 vvm; 4, Q_G = 1.04 vvm; 5, Q_G = 1.32 vvm).

Increases in gassing rate just alter the frequency with which gas is released from the cavities. The same phenomenon occurs at the higher speeds discussed below.

Over the speed range giving rise to Reynolds numbers up to about 300 to 400, the power number falls due to a combination of the streamlining effect, the reduced pumping capacity of the impeller and the increased pressure in the gas-filled cavities. At this point, a minimum power number is reached. The cavities are of maximum size and, in addition, the impeller is rotating in what appears to be a pocket of gas from which little actual dispersion occurs into the rest of the vessel[20].

For increases in speed and Reynolds number beyond those at

which this minimum power number occurs (shown for a variety of fluids for disc turbines in Fig. 7), the size of the cavities begins to decrease and the power number increases again. In this region, $10 \le Re \le 1000$, there is no effect of gassing rate if the fluid is significantly viscoelastic (i.e., the CMC and Boger fluids of Table 1) and only a weak effect even if it is inelastic[9, 14].

As the viscosity of the fluids falls, so the Reynolds number at which the minimum occurs increases and the dependency of power number on gassing rate approaches that found with fluids at Reynolds numbers larger than 1000.

The actual value of the minimum gassed power number, $(Po_g)_{min}$, depends in a complex way on the size of the cavities, on the amount of gas held in the impeller region and on the pumping capacity. Caverns can form in aerated fluids[16] just as they can in unaerated ones as described in Chapter 4. As a result, the gas held in the region of the impeller is greatest if the fluid has a yield stress or is extremely shear-thinning. Pumping capacity is also reduced by increasing viscoelasticity, yield stress or pseudoplasticity.

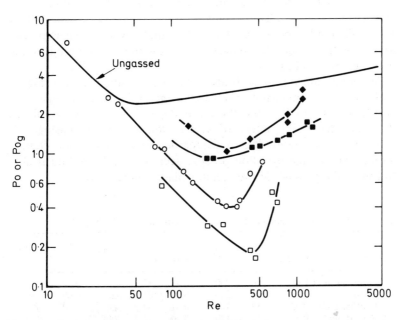

Figure 7   Gassed power number versus Reynolds number at $Q_G = 0.5$ and 1 vvm: $10 \le Re \le 1100$[13] (geometry, as Fig. 6; 0, 1.4% CMC; □, 4.5% Xanthan gum; ■, 2% Xanthan; ◆, 0.17% Carbopol).

Attempts to correlate the value of $(P_g/P)_{min}$ with rheological properties and agitation conditions have been made using the dimensionless groups discussed in Chapter 4. The Weissenberg number is considered to relate viscous and elastic forces and is defined as

$$\text{Wi} = \nu_1/\tau = (A/K)(\dot{\gamma}_{AV})^{b-n} \tag{8}$$

if it is assumed that the average shear rate determined by the method of Metzner and Otto[4] can be applied to parameters other than viscosity. Alternatively, the ratio of elastic to inertial forces can be derived similarly as

$$m = \text{Wi}/\text{Re} = (A/\rho ND^2)(\dot{\gamma}_{AV})^{b-1}. \tag{9}$$

Table 3 lists all the above parameters for a number of the fluids of Table 1 at $\dot{\gamma}_{AV}$ derived from the Reynolds number (or agitation speed) at which the minimum value of Po_g was found[13]. Though none of the parameters correlate very well with the reduction in P_g/P, the first normal stress difference, ν_1, a direct measure of the elasticity, is the best. This is shown in Fig. 8. Thus it is seen that, at Reynolds number between 10 and 1000, high levels of elasticity lead to a reduction in aerated power. This reduction is enhanced by the presence of a yield stress.

The minimum Po_g value is also a function of impeller size (Fig. 9) and scale of operation. In both cases, increasing size increases the absolute value of Po_g and therefore Po_g/Po^9. It can be postulated that this decreasing effect of viscoelasticity with increasing size actually comes about due to decreasing speed. Thus, the characteristic time of the agitation process decreases and the fluid is able to follow the passage of the agitator blades more readily.

Table 3   Fluid Parameters and Dimensionless Groups and $(P_g/P)_{min}$ [13]

Fluid	$(P_g/P)_{min}$	N_{min} s^{-1}	$\dot{\gamma}_{AV}$ s^{-1}	μ_a Pas	ν_1 Pa	Re_{min} —	Wi —	$m \times 10^3$ —
4.3% Xanthan gum	0.07	20	230	0.45	82	420	0.80	1.9
1.4% CMC	0.14	18	210	0.59	102	320	0.69	2.2
2% Xanthan gum	0.35	12	140	0.31	34.5	300	0.79	2.6
0.17% Carbopol	0.40	11	130	0.30	13.2	300	0.27	0.7
0.3% Carbopol	0.38	15	170	0.60	35.4	250	0.32	1.3

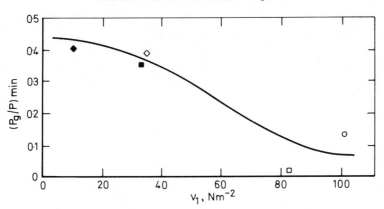

Figure 8   The minimum gassed to ungassed power ratio as a function for first normal stress difference[13]. (Geometry as Fig. 6; Symbols as Fig. 7 plus $\diamond$, 0.3% Carbopol).

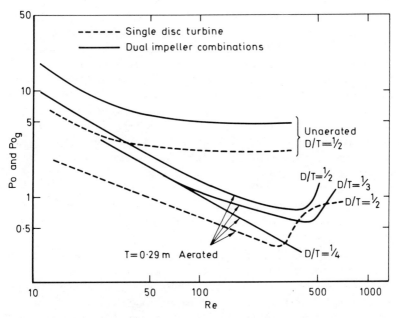

Figure 9   Gassed and ungassed power numbers for single and dual impellers (disc turbine and angled blade) in 4.5% Xanthan gum[1].

7. IMPELLER COMBINATIONS FOR GOOD MIXING AND AERATION

An estimate of the agitator speed required for gas dispersion can be obtained from the previous section and Chapter 5. However, a single agitator will not usually enable good overall mixing to be obtained in the Reynolds number region less than 1000. To do so, it is necessary to use more than one impeller as explained in an earlier section[1].

Chapter 4 showed clearly the complexity of the interaction that can occur between agitator type and rheological properties. This complexity is enhanced for multiple impeller systems where different types, sizes and positioning of impeller may occur. The addition of gas adds further to the difficulty of making a precise analysis of the problem. Yet these configurations are regularly used in industry, especially in the aeration of rheologically-complex fermentation fluids. A few general comments will have to suffice here.

First, if driven from the same shaft, the use of different sized agitators must be used with caution. Unless the differences are very small, the larger agitator draws by far the greatest power, even for very different power numbers. The lower agitator acts as the gas disperser and experiences a significant drop in power compared to the ungassed condition, as outlined in the previous section. However, this fall in power is less if relatively large D/T ratios are used (see Fig. 9). The upper agitator mainly acts as a device for aiding turnover of the tank contents. Because of the low sensitivity of the cavity structure to gassing rate, the drop in the power drawn is similar to that which would be found if the gas was sparged directly into it[9]. In this respect, the upper impeller behaves very differently as compared to systems with low viscosity liquids where it draws almost as much power as when ungassed[21]. Therefore, the power drawn by a dual impeller combination falls from the ungassed value almost as much as it does for a single impeller (Fig. 9)[1]. Thus, from the point of view of power drawn, relatively large, for example, $D/T = 0.5$, impeller combinations are recommended.

As explained in Chapter 5 in highly shear-thinning fluids, agitators can form caverns. This is, of course, also true when agitator combinations are used. If the upper, "mixing" agitator is an angled blade agitator, then the flow pattern within the cavern depends on whether the angled blade is pumping upwards (Fig. 10a) or downwards (Figs. 10b or 10c)[16]. Though large impeller combinations give fluid movement throughout the vessel at acceptable power levels[22],

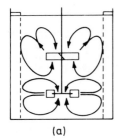 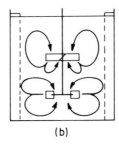 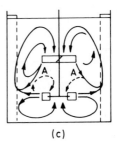

(a)   (b)   (c)

Figure 10   Flow patterns in unaerated high viscosity, shear–thinning fluids agitated by a dual impeller combination[16]. (a) The angled blade agitator pumping upwards; (b) The angled blade agitator pumping downwards at low speeds; (c) The angled blade agitator pumping downwards at high speeds.

top-to-bottom mixing is very poor when the angled blade is pumping upward[23].

The addition of gas does not alter the basic flow pattern, though it may damp out the fluid movement, relative to the un-gassed condition. Gas dispersion is maintained within the cavern except for the occasional large bubble which coalesces and breaks away from it. Until the cavern extends to the wall, the tank cannot be considered well-mixed nor the gas well-dispersed. It has been shown[1] that, of the two modes of agitation, the downward pumping mode at sufficiently high speeds to give rise to the flow pattern of Fig. 10c is best for overall gas dispersion. Under these conditions, the severe shearing in region A gives an additional break-up action. This shearing reduces the tendency for the large gas bubbles leaving the gas-filled cavities on the back of the lower impeller to pass through without further disintegration. On the other hand, the upward-pumping mode gives a large cavern at the same speed.

A force-balance between the torque on the cavern surface and that on the agitator has enabled the speed of agitation to be linked to the cavern size[22]. For dual impellers, the equation

$$\log (r/D) = 1/3 \log (\mathrm{Po}_g N^2) + 1/3 \log (\rho D^2/4\pi^3 \tau'_y) \qquad (10)$$

where r is the radius of the cavern measured from the centre of the nearest impeller. τ'_y is either the yield stress or the shear stress at an acceptable minimum shear rate value, that is, a rate that represents the minimum turnover of fluid for the process in question. τ'_y can be obtained from the flow curve.

The power required for complete mixing drops very rapidly with increasing size of impeller for both gassed and ungassed mixing.

Table 4   Comparison of Power Requirements for Complete Mixing
(2% Xanthan gum)[22]

Power input	Impeller combinations				Disc turbine
	0.6 T	0.5 T	0.33 T	0.25 T	0.5 T
ϵ_T(W/kg)	1.2	2.2	3.2	4.9	15.7

Table 4 shows data for the ungassed case and the inadequacy of the
single agitator is very obvious. Overall, dual impeller combinations
of large diameter are essential for adequate gas dispersion and
mixing in high viscosity rheologically complex fluids.

8.   MASS TRANSFER RATES AND HOLD-UP

These two aspects of gas–liquid mixing are very commonly studied.
However, rarely are the difficulties involved in measuring them
properly reported. It is important to emphasise that both are
extremely sensitive to trace quantities of surface active contaminants
in solution or solids in suspension in the liquid. In addition, mass
transfer rates in terms of $k_L a$ values, are very closely related to
hold-up. This is because both are global averages, though hold-up
values have been shown to vary markedly from place to place[24] and $k_L a$
almost certainly does too.

$$\epsilon = \Delta h/(H + \Delta h) \tag{11}$$

and is an average for the whole of the fluid in the vessel. Mass
transfer rates are usually expressed as $k_L a$ because of the difficulty
of separating k_L, the liquid film mass transfer coefficient and, a, the
area of gas–liquid contact/unit volume of fluid. If the mean bubble
size is d_B then,

$$d_B = 6\epsilon/a \tag{12}$$

provided that the bubbles are very nearly spherical.

For a particular fluid, d_B is not greatly affected by the level of
agitation and neither is k_L. Increasing agitation speed and power
enhances ϵ by increasing the amount of gas recirculation and, by the
same means, enhances a and thus $k_L a$.

8.1.   Hold-Up

In a batch fermentation (or polymerization) reactor, the range of
bubble sizes is, initially, relatively narrow. If the fluids are non-
coalescing, for example, electrolyte solutions, small bubbles form and

if coalescing (pure liquids), relatively large ones[25]. Thus, non-coalescing systems have higher hold-ups than coalescing. As the reaction proceeds and the liquid becomes more viscous and slightly non-Newtonian, the bubble size distribution becomes bi-modal with a large number of very small bubbles and a small number of very large bubbles[26]. As a result, for the same agitation conditions, the larger bubbles have a shorter residence time and the hold-up decreases. The absolute hold-up still depends on whether impurities are present thereby reducing the "small-bubble" coalescence[27].

As the viscosity of the non-Newtonian fluid increases further, break-up of bubbles by the agitator becomes progressively more difficult. In addition, coalescence occurs more and more readily. Thus, some extremely large bubbles with a very short residence time pass through. These large bubbles are sherical cap ones, though, if the fluid is viscoelastic, they are found to have a characteristic tail at the rear. At the same time, the very tiny bubbles become more or less indistinguishable from the fluid itself, recirculate many times and, if mass transfer is occurring, quickly become depleted of the transferring substance[28].

As implied earlier, the accurate measurement of hold-up in high viscosity fluids is difficult and the method suggested in Chapter 5 cannot be used. A very time-consuming method has been described[26] and, as such, its use cannot generally be justified. A simple visual assessment of the enhanced liquid level works quite well on the small scale, especially with viscous fluids when the upper interface is fairly quiescent. The eruption of large bubbles and foaming are the main difficulties.

Allowing for the difficulties in determining ϵ, Fig. 11 shows the

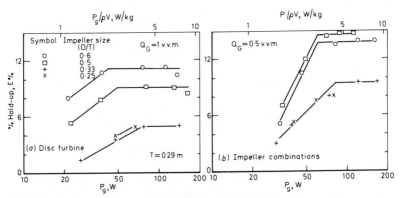

Figure 11 Examples of the relationship between hold-up and power input when dispersing gas in a high viscosity, shear-thinning fluid[1]. (a) A single disc turbine; (b) A combination of 45° downward pumping impeller and disc turbine.

change of hold-up with increasing power input[29] for Xanthan gum solutions for both single disc turbine impellers and for impeller combinations. As can be seen, above a certain power, no further increase in hold-up occurs. Larger impellers give this effect at lower power inputs and, in addition, for a given power input, large diameter combinations of impeller (disc turbine and 45° pitched blade) give the highest hold-up.

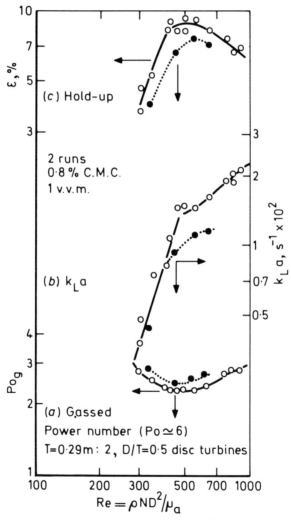

Figure 12 The interaction between power, hold-up and $k_L a$ in high viscosity shear-thinning fluids[51]. (a) Po_g vs. Re; (b) $k_L a$ vs. Re; (c) ϵ vs. Re.

More recent work[51] has clearly shown the link between hold-up, cavity formation and power drawn (Fig. 12)[29]. The figure shows that the minimum of the gassed power curve (Fig. 12a) (which corresponds to the largest gas-filled cavity size and bubble cloud in the impeller region) corresponds to the maximum hold-up (Fig. 12c). At higher speeds (expressed as Reynolds number), as this local concentration of gas is dispersed, the overall hold-up falls even though the hold-up away from the impeller region would be expected to continue to increase.

8.2. Diffusion Coefficients in Solutions of Macromolecular Media

Diffusion coefficient, like the density and the viscosity of a liquid, is one of the key correlating parameters in any design equation for mass transfer operations and, thus, its determination is of some importance. Since the rate of momentum transport in non-Newtonian liquids is a non-linear function of the state of stress (or of the rate of deformation) one is naturally concerned about the diffusion coefficient being also dependent on the deformation rate in the flowing liquid. Further, the diffusion coefficient of molecules in Newtonian media is an inverse function of viscosity, known as the Stokes–Einstein equation

$$\mathscr{D}\mu/\theta = \text{const} . \tag{13}$$

As of today it is not quite clear how this formula should be modified to hold for non-Newtonian polymer solutions, the viscosity of which is a function of shear rate.

Even though the static techniques for the measurement of diffusion coefficients do not allow for a controlled variation of shear rate, the dynamic techniques are also not without drawbacks. This is because the flow field must be known (and well defined) in order to back-calculate the diffusion coefficient from the measured integral mass transfer rate. With non-Newtonian polymer solutions, particularly if they are viscoelastic, this is not always the case.

Most experimental techniques currently used to determine the diffusion coefficient of solutes in polymer solutions will fall in one of the following four categories: absorption of gases into jets or films, electrochemically aided mass transport, and laminar dispersion in tubular flows. Each of these four techniques suffers from specific drawbacks which bring about specific experimental errors: measuring of small mass differences or of small concentration differences, modification of the rheological properties of the solutions used by ionic additives, and the adoption of certain unrealistic assumptions about the velocities and their gradients.

There is a great deal of confusion about the influence of polymer concentration, the shear rate, and the temperature on the diffusion coefficient of solutes in polymer solutions. It would seem, when examining the available data critically, that the experimental results depend, to a large extent, on the experimental technique used.

Osmers and Metzner[30] used an optical wedge to measure the diffusivity of glycerol, ethanol, and allylalcohol in aqueous solutions of carboxymethylcellulose (CMC) and polyacrylamide (PAA). They found that the diffusion coefficient decreases as the polymer concentration increases but this reduction of $\mathscr{D}$ is at least one order of magnitude less than the increment in the solution's zero-shear viscosity. Specifically, in solutions containing 10% of polymer by weight the diffusion coefficient of the three solutes mentioned above is only 20% lower than in pure water although the zero-shear viscosity is at least two orders of magnitude above that for pure water.

Using a technique of a laminar liquid jet issued into a gaseous environment, Astarita[31], Zandi and Turner[32], and Dim et al.[33] found no clear trend in the diffusion coefficient of carbon dioxide in aqueous solutions of Carbopol, Methocel, hydroxyethylcellulose (HEC), CMC, and PAA when the polymer concentration was varied. It must be realized, however, that most of the polymer solutions cited above are not only shear-thinning but also visco-elastic in the concentrations used. Therefore, the shape of such a liquid jet might not be as one would expect it for an inelastic liquid. This might account for an initially slight increase followed by a decrease of the diffusion coefficient vs. concentration curve.

A non-rippling film of a non-Newtonian liquid flowing down an inclined or a vertical surface is probably a safer geometrical arrangement to be used with polymer solutions. A falling film technique was used by Wasan et al.[34] and by Perez and Sandall[35] to measure the diffusion coefficient of oxygen and carbon dioxide in aqueous solutions of Carbopol, Polyox and CMC. Perez and Sandall used a cylindrical column about six times longer than that used by Wasan et al. and, therefore, their results are more representative. They found that by dissolving 1% of Carbopol in water the diffusion coefficient of carbon dioxide drops from $2.0 \times 10^{-5}\,\mathrm{cm^2/s}$ to about $1.6 \times 10^{-5}\,\mathrm{cm^2/s}$. The integral mass transfer data were processed using the penetration theory which is probably an adequate approach considering that Carbopol solutions at 1% concentration might display a yield stress. The falling film technique is not particularly suitable to study the influence of shear rate since the velocity profile in the film at the gas-liquid interface is essentially flat. This must be borne in mind when considering a slight increase of the diffusion coefficient with the increase of the flow rate found by Wasan et al.[34]

Also Mashelkar et al.[36, 37] employed the falling film technique but rather than the cylindrical walls of a column they used a sphere[36] and a cone[37] to study the diffusion coefficient of carbon dioxide in aqueous solutions of Polyox, CMC, PAA, and HEC. In this geometry, the liquid in the film undergoes not only shearing, but also stretching deformation as it accelerates from the pole (or the apex): the authors argue that the relaxation times of the liquids used were such that the liquids could accommodate the strain. Using this technique they found that the diffusion coefficient, before it starts decreasing with the increasing concentration of the polymer, will show some dramatic increase. In the case of a 0.5% CMC solution, the increase was found to be 40%.

Using a spinning disk electrode, Greiff et al.[38, 39] found a 40% reduction of the diffusion coefficient of oxygen through a saline solution of 1% of Polyox. Greiff et al. varied the average shear rate by a factor of one hundred. By inspecting their data no influence of the shear rate on the diffusion coefficient can be detected. Unfortunately, Greiff et al. do not give any account of the secondary flow around the spinning disk so that it may be suspected that at least a part of this reduction is due to the suppression of the radial flow (for details see Chapter 4).

When the apparent (shear-dependent) viscosity μ_a is used in Eq. (13) rather than the Newtonian viscosity μ, the variation of the Stokes-Einstein coefficient is not entirely eliminated[40]. It would seem, therefore, that the apparent viscosity is not in control of the diffusion coefficient. On the contrary, Arvia et al[41], working with a rotating disk electrode, found in some cases $D\eta/T$ independent of the shear rate which would imply the diffusion coefficient increasing with increasing shear rate (a result reported also by Mashelkar and Soylu[36]).

It has been known for some time that the rheological properties of most polymer solutions may be drastically changed by the presence of ions. Since the electrochemical methods of measurement of diffusion coefficients require a fairly high concentration of a strong electrolyte in order to maintain the electric current at a measurable level, there must be some uncertainty about the propriety of the electrochemical methods used.

The account of experimental techniques used or designed for use with non-Newtonian polymer solutions would be incomplete without mentioning the method based on the laminar tubular dispersion of a solute from a point source. A complete solution to this problem for a power-law and Ellis liquid was presented by Venkatsubramanian et al.[42] who also provided experimental data obtained with NaOH dispersed in Natrasol. The results of both the theoretical and the experimental analysis show that the shear rate has no significant influence on the value of the diffusion coefficient and that the influence of the polymer concentration is relatively weak.

In conclusion, the general consensus seems to be that the diffusion coefficient of low molecular solutes in non-Newtonian polymer solutions will not greatly differ from that in pure solvent. At high concentrations of dissolved polymer, reduction of the diffusion coefficient around 20% can be expected. Navari et al.,[43] however, report reductions around 50% to 60% for the diffusion of oxygen in plasma which has been "thickened" by the addition of CMC.

The evidence regarding the influence of the shear rate on the diffusion coefficient of solutes has been conflicting so far. In any case, however, the deformation rate is not likely to play a strong role.

8.3.   The Overall Mass Transfer Coefficient

The overall liquid-side mass transfer coefficient "$k_L a$" has been the subject of interest for some time. Loucaides and McManamey[44] studied the oxygen transport to a simulated fermentation broth (paper pulp suspension) but they did not include the rheological parameters in their final correlation.

Perez and Sandall[45] measured gas absorption in polymer solutions stirred by a turbine and concluded that the modified Sherwood number

$$Sh' = k_L a d_B^2 / \mathcal{D} = Sh \cdot a d_B$$

correlates well with

$$Sc' = \mu_a / \rho \mathcal{D}$$

and

$$Re' = \rho d_B^2 N / \mu_a$$

where the apparent viscosity was taken as

$$\mu_a = K(11N)^{n-1}[(3n+1)/4n]^n$$

In their final correlation

$$Sh' = B(Sc')^{0.5}(Re')^{1.11} \tag{14}$$

the parameter B incorporates also the influence of interfacial tension but it is not dimensionless which makes the use of Eq. (14) difficult.

The influence of the polymer concentration on the mass transfer coefficient $k_L a$ in stirred tanks was systematically investigated by Konig et al[46]. They found that the coefficient always decreases when the polymer concentration increases. On the other hand, the coefficient may either increase or decrease with the increase of the gas flow rate depending on the concentration of the polymer.

The coefficient $k_L a$ can be determined in a number of ways but none of them is without problems. In low viscosity fluids, the dynamic method is very convenient. However, if the oxygen concentration in the gas phase falls to a low level then $k_L a$ determined from one experiment can vary by a factor of up to about 3 depending on the gas mixing pattern assumed[47, 48]. As the $k_L a$ value falls, the assumption made about the gas phase mixing becomes less important.

However, if the fall in $k_L a$ comes about because of increasing viscosity and non-Newtonian behaviour, the assumption of a well-mixed liquid phase becomes less valid. This is a particularly difficult problem to detect using the unsteady state method. Other problems also arise when using this method. When degassing with nitrogen, a permanent hold-up of tiny N_2 bubbles remains after degassing and before aeration. These bubbles give rise to low values of $k_L a$[49]. The response of the oxygen electrode needs consideration. Though fast oxygen electrodes have largely obviated the need to make allowance for their response time in low viscosity fluids, this is not true in high viscosity ones. In this case, the fluid causes a boundary layer effect so that the "probe plus fluid" response time becomes very significant and a function of the agitation speed. Kipke[18] got round the problem by using a small second agitator close to the electrode, thus giving a response time similar to that in low viscosity fluids.

The steady state technique, based on a mass balance to give the mass flux, overcomes many of these problems. However, the absolute solubility of the oxygen in the fluid is required and the use of the liquid phase for the mass balance causes problems because it is difficult to achieve liquid throughflow. Basing the mass balance on the gas phase requires a suitable oxygen sink. Chemical methods such as sulphite oxidation and the bicarbonate–carbonate reaction are not suitable because they radically alter the rheological properties of the solution. Indeed, in the work by Ranade and Ulbrecht[50], the viscosity was reduced to a value close to that of water.

The use of a respiring organism gives a reasonable system and yeast has been used with some success. The $k_L a$ value still changes somewhat with time during the course of a run because of the metabolites produced. However, this effect can be allowed for by only comparing results from the same time of respiration. The use of two or three oxygen electrodes enable a measure of the quality of the liquid phase mixing to be assessed. Visual observation is also a help. The gas phase mixing may not be known but the $k_L a$ values are low so that the assumption of perfect mixing or plug flow makes little difference[51].

Figure 12b shows how $k_L a$ increases rapidly with increasing speed (Reynolds number) up to the minimum in Po_g and the maximum in the hold-up. After this, it increases more slowly. It would be wrong, just because the hold-up falls, to ascribe the increase in $k_L a$ to increases in k_L[51]. This is because though the overall hold-up falls, the average value of a may still increase as a result of the enormous differences in bubble size associated with regions with different hold-ups; the very large cavities and bubble cloud in the region of the impellers[16]; the large coalesced bubbles of short residence time; and the tiny, long residence time recirculating bubbles.

Figure 13 shows the broad trend relating $k_L a$ to apparent viscosity at an energy dissipation rate of $2.5 \, kW/m^3$. The effect of aeration rate is very small at the higher viscosities and no attempt is made here to separate results from 1/2 and 1 vvm aeration rates. The earlier sections on hydrodynamics show the relative unimportance of aeration rate in these systems, so that the small effect on $k_L a$ is hardly surprising.

Further, Fig. 14 combines together the recent data of Hickman[51] and some earlier data of Kipke[18] and Solomon[1] for high viscosity fluids. Also included are low viscosity values from Warmoeskerken and Smith[52]. (This latter data is also presented in Figure 76 of Chapter 5). The correlation technique is one recommended by Zlokarnik[53]. At first sight, the plot correlates all the data. However, it is important to recognise that, at the same values of $(P_g/Q_G)^*$, $k_L a$ values are varying by a factor of two in places. As with all log–log plots covering many orders of magnitude, it should be used with caution.

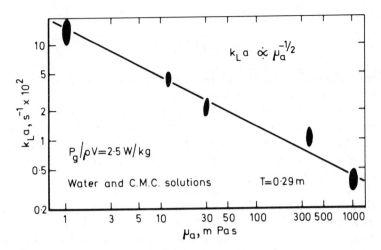

Figure 13   The effect of viscosity on k_La using water and a range of C.M.C. solutions[29].

The change in slope indicated by Fig. 14 between low and high viscosity fluids can be interpreted as a reduction in the effect of gassing rate and an increase in the effect or power input. For the high viscosity range,

$$k_La \propto Q_G^{0.34}(\epsilon_T)_g^{0.66} \tag{15}$$

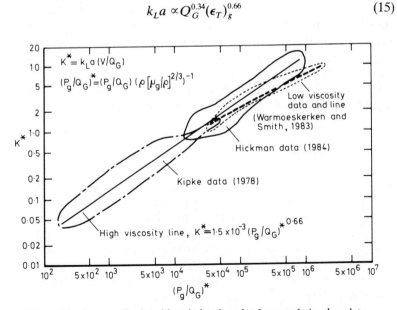

Figure 14   A generalised multi-cycle log–log plot for correlating k_La data.

whilst for the low viscosity

$$k_L a \propto Q_G^{0.55}(\epsilon_T)_g^{0.45} \tag{16}$$

This reduced sensitivity to gassing rate has important implications for the control of fermentation processes[29].

Finally, we need to mention briefly the influence of viscoelasticity on the mass transfer coefficient $k_L a$. Yagi and Yoshida[20] put forward a correlation for the modified Sherwood number Sh but they characterized the viscoelasticity by a time parameter derived from a steady state flow curve.

In a paper already mentioned in this chapter, Ranade and Ulbrecht[15] compared the overall mass transfer coefficients in inelastic CMC and viscoelastic PAA solutions. Although the viscosities of these two solutions were significantly reduced by the presence of ions, the difference of their elastic behaviour was not affected to the same extent. It was found that the value of $k_L a$ in the PAA solutions was lower than in the CMC solutions by a factor of two and more although the viscosities of the solutions were equal. Following the arguments of Kolmogoroff, the reduction of the true liquid side mass transfer coefficient k_L should be proportional to the fourth root of the reduction of power consumption due to the addition of the polyacrylamide. Since the maximum reduction of power is about 70%, the mass transfer coefficient should not be reduced by more than about 15%. Any further reduction must be, therefore, on the account of the reduction of the specific interfacial area a. This is consistent with the fact that viscoelastic liquids show an increased resistance to stretching. Thus, a large bubble which would be easily broken into several small bubbles in an inelastic solution, will retain its large size in a viscoelastic liquid.

9. CONCLUSIONS

Chapter 4 shows the complex flow patterns that occur in mixing vessels when fluids of different rheological properties are treated within them. These flow pattern complexities are compounded when the question is posed "How do these complexities affect either the ability of the mixer to mix or the quality of the mixture obtained". Chapter 5 shows that, even in low viscosity Newtonian fluids, the dispersion of gas depends very markedly on the agitator type and on the gassing rate.

In this chapter, an attempt has been made to clarify the real industrial problems which involve both complex rheology fluids and gas dispersion into them. It has highlighted the rheological properties which most affect gas dispersion mechanisms, agitator power drawn, overall liquid and gas–liquid circulation and gas–liquid mass transfer rates. This chapter more perhaps than any other in this book needs to be read in close conjunction with other ones, that is to say, Chapters 4 and 5 particularly and also some parts of Chapter 8. Though good progress is being made, much more work still needs to be done before gas–liquid dispersion in high viscosity systems is well understood.

10. NOTATION

a specific surface area of gas bubbles, m^{-1}
A constant in Eq. 4, $N\,s^b m^{-2}$
b constant in Eq. 4, dimensionless
d_B bubble diameter, m
D impeller diameter, m
$\mathscr{D}$ diffusivity of gas phase through liquid, $m^2\,s^{-1}$
Fl_G gas flow number $= Q_G/ND^3$, dimensionless
Fr Froude number $= N^2D/g$, dimensionless
H vessel unaerated liquid height, m
K^* reduced $k_L a$ (see Fig. 14), dimensionless
K consistency index, $N\,s^n m^{-2}$
k_s impeller shear rate constant $(j_{AV} = k_s N)$, dimensionless
$k_L a$ mass transfer coefficient, s^{-1}
m Wi/Re, dimensionless
n flow behaviour index, dimensionless
N agitator speed, s^{-1}
P power drawn in liquid, W
Po power number $= P/\rho N^3 D^5$, dimensionless
$(P_g/Q_G)^*$ reduced power number (see Fig. 14), dimensionless
Q_G sparged gas rate, $m^3\,s^{-1}$
r cavern radius, m
Re' Reynolds number $= \rho d_B^2 N/\mu_a$, dimensionless
Re Reynolds number $= ND^2/\mu_a$, dimensionless
Sc' Schmidt number $= \mu_a/\rho\mathscr{D}$
Sh Sherwood number $= k_L d_g/\mathscr{D}$, dimensionless
Sh' modified Sherwood number (see Section 8.3), dimensionless

T vessel diameter, m
V volume of liquid in vessel, m^3
Wi ν_1/τ, dimensionless

Greek Symbols

Δh increase in liquid height due to gassing, m
ϵ hold-up, dimensionless
ϵ_T power per unit mass, $W\,kg^{-1}$
$\dot{\gamma}$ shear rate, s^{-1}
μ_a apparent dynamic viscosity, $N\,s\,m^{-2\,1}$
μ dynamic viscosity, $N\,s\,m^{-2}$
ν_1 1st normal stress difference, $N\,m^{-2}$
ρ liquid density, $kg\,m^{-3}$
τ shear stress, $N\,m^{-2}$
τ_y yield stress, $N\,m^{-2}$
τ'_y apparent yield stress (see Eq. 10), $N\,m^{-2}$
θ absolute temperature, K

Subscripts

g gassed
AV   at the average shear rate (see Eq. 4)
min   minimum
CD   when the gas is just completely dispersed
HB   Herschel–Bulkley parameters (see Eq. 1)
C Casson parameters (see Eq. 3)

REFERENCES

1. Solomon, J. (1980). Mixing, aeration and rheology of highly viscous fluids, Ph.D. Thesis, University of London.
2. Metz, B. (1976). From pulp to pellet, Ph.D. Thesis, University of Delft.
3. Boger, D.V. (1977/78). A highly elastic constant viscosity fluid, *J. Non-Newt. Fluid Mech.*, **3**, 87.
4. Metzner, A. B., and Otto, R. E. (1957). Agitation of non-Newtonian fluids, *AIChE J.*, **3**, 3.
5. Rushton, J. M., and Oldshue, J. Y. (1953). Mixing-present theory and practice, *Chem. Eng. Prog.*, **49**, 161 *and* 267.
6. Anderson, C., LeGrys, G. A., and Solomons, G. L. (1982). Concepts in the design of large scale fermentors for viscous culture broths, *The Chem. Eng.*, Feb., 43.

7. Solomon, J., Nienow, A. W., and Pace. G. W. (1981a). Power and hold-up in the mixing of aerated, viscous non-Newtonian fluids, *Adv. in Biotechnol.*—Vol. I, Scientific and engineering principles (*Proc. 6th Int. Fermentation Symp.*), Pergamon Press, pp. 503–509.
8. Kuboi, R., Nienow, A. W., and Allsford, K. V. (1983). A multipurpose stirred tank facility for flow visualisation and dual impeller power measurement, *Chem. Eng. Commun.*, **22**, 29.
9. Allsford, K. V. (1985). Gas–liquid dispersion and mixing in mechanically agitated vessels with a range of fluids. Ph.D. Thesis, University of Birmingham, UK.
10. Van't Riet, K. (1975). Turbine agitator hydrodynamics and dispersion performance, Ph.D. Thesis, University of Delft.
11. Nienow, A. W., Allsford, K. V., and Kuboi, R. (1983a). Using a derotational prism for studying gas dispersion processes, *Paper to Eng. Found. Conf. on Mixing*, Henniker, New York.
12. Nienow, A. W., Warmoeskerken, M. M. C. G., Smith, J. M., and Konno, M. (1985). On the flooding/loading transition and the complete disposal condition in aerated vessels agitated by a Rushton turbine, *Proc. 5th European Mixing Conf.*, BHRA Fluid Engineering, Cranfield, Bedford, England, pp. 143–154.
13. Nienow, A. W., Wisdom, D. J., Solomon, J., Machon, V., and Vlcek, J. (1983c). The effect of rheological complexities on power consumption in an aerated agitated vessel, *Chem. Eng. Commun.*, **19**, 273.
14. Nienow, A. W., Allsford, K. V., and Kuboi, R. (1983b). The effect of rheological properties on gas-filled cavity formation when dispersing gas with rotating stirrers at Reynolds number < 1000, *Paper to Conf. on Eng. Rheology*, London, England.
15. Ranade, V. R., and Ulbrecht, J. J. (1977). Gas dispersion in agitated inelastic and viscoelastic liquids, *Proc. 2nd European Mixing Conf.*, BHRA Fluid Engineering, pp. F6-83 to F6-89.
16. Solomon, J., Nienow, A. W., and Pace, G. W. (1981b). Flow patterns in agitated plastic and pseudoplastic viscoelastic fluids, in *Fluid Mixing*, I. Chem. E. Symp. Ser., No. 64, pp. A1–A13.
17. Nienow, A. W., Kuboi, R., Chapman, C. M., and Allsford, K. V. (1983d). The dispersion of gases into liquids by mixed flow agitators, *Proc. Int. Conf. on Physical Modelling of Multiphase Flows*, BHRA Fluid Engineering, Cranfield, Bedford, England, pp. 417–438.
18. Kipke, K. D. (1978). Gas dispersion in non-Newtonian liquids, *Proc. Int. Symp. on Mixing*, Fac. Poly de Mons, pp. C5–1 to C5–21 and XC60 to XC70.
19. Greaves, M., and Loh, V. Y. (1984). Power consumption effect in three-phase mixing, in *Fluid Mixing II*, I. Chem. E. Symp. Ses., No. 89, pp. 69–96.
20. Yagi, H. and Yoshida, F. (1975). Gas absorption by Newtonian and non-Newtonian fluids in sparged, agitated vessels, *Ind. Eng. Chem. Proc. Des. Dev.*, **14**, 488.
21. Kuboi, R., and Nienow, A. W. (1982). The power drawn by dual impeller systems under gassed and ungassed conditions, *Proc. 4th European Mixing Conf.*, BHRA Fluid Engineering, Cranfield, Bedford, England, pp. 247–261.
22. Solomon, J., Elson, T. P., Nienow, A. W., and Pace, G. W. (1981c). Cavern sizes in agitated fluids with a yield stress, **11**, 143.
23. Nienow, A. W., and Kuboi, R. (1984). A technique for studying intervortex mixing rates in a dual impeller agitated vessel in high viscosity fluids, in *Fluid Mixing II*, I. Chem. E. Symp. Ser., No. 89, pp. 97–106.
24. Nienow, A. W., Wisdom, D. J., and Middleton, J. C. (1978). The effect of scale and geometry in flooding, recirculation and power in gassed, stirred vessels, *Proc. 2nd European Mixing Conf.*, BHRA Fluid Engineering, Cranfield, Bedford, England, pp. F1–1 to F1–16 and pp. X52 to X54.
25. Moo-Young, M., and Blanch, H. W. (1981). Design of biochemical reactors, *Adv. Biochem. Eng.* **19**, 1.

26. Machon, V., Vlcek, J., Nienow, A. W., and Solomon, J. (1980). Some effects of pseudoplasticity on hold-up in aerated, agitated vessels, *Chem. Eng. J.*, **14**, 67.
27. Nienow, A. W., and Machon, V. (1979). The effect of formaldehyde on hold-up in aerated, stirred tanks, *Biotechol. Bioeng.*, **21**, 1477.
28. Schugerl, K. (1981). Oxygen transfer into highly viscous media, *Adv. Biochem. Eng.*, **19**, 72.
29. Nienow, A. W. (1984). Mixing studies on high viscosity fermentation processes–Xanthan gum, *World Biotechnology Report*, Vol. 1 Europe. Online (London), pp. 293–304.
30. Osmers, H. R., and Metzner, A. B. (1972). Diffusion in dilute polymeric solutions *Ind. Eng. Chem., Fundam.*, **11**, 161.
31. Astarita, G., and Marrucci, G. (1965). Diffusivity in non-Newtonian liquids, *Ind. Eng. Chem., Fundam.*, **4**, 236.
32. Zandi, I., and Turner, C. D. (1970). Absorption of oxygen by dilute polymeric solutions-molecular diffusivity measurement, *Chem. Eng. Sci.*, **25**, 517.
33. Dim, A., and Ponter, A. B. (1971). Absorption of oxygen by dilute polymeric solutions: molecular diffusivity measurements, *Chem. Eng. Sci.*, **26**, 1301.
34. Wasan, D. T., Lynch, M. A., Chad, K. J., and Srinivas, N. (1971). Mass transfer into dilute polymeric solutions, *AIChE J.*, **17**, 1028.
35. Perez, J. F., and Sandall, O. C. (1973). Diffusivity measurement for gases in power-law non-Newtonian liquids, *AIChE J.*, **19**, 1073.
36. Masheklar, R. A., and Soylu, M. A. (1974). Diffusion in flowing films of dilute polymeric solutions, *Chem. Eng. Sci.*, **29**, 1089.
37. Mashelkar, R. A., and Soylu, M. A. (1982). Gas diffusion in polymer solutions: A double-cone flow technique, *J. Appl. Polym. Sci.*, **27**, 697.
38. Greif, R., Kapesser, and Cornet, I. (1972). Transport in non-Newtonian flow, *J. Elchem. Soc.*, **119**, 717.
39. Greif, R., and Patterson, J. A. (1973). Mass transfer to a rotating disk in a non-Newtonian fluid, *Phys. Fluids* **16**, 1816.
40. Klinger-Park, P. U., and Hubbard, D. W. (1985). Diffusion in non-ionic and ionic polymer solutions: Effect of shear rate, *Chem. Eng. Commun.*, **32**, 171.
41. Arvia, A. J., Bazan, J. C., and Carozza, J. S. (1968). Diffusion of ferro- and ferricyanide ions in aqueous potassium chloride solutions and in solutions containing carboxymethylcellulose sodium salt, *Electrochem. Acta*, **13**, 81.
42. Venkatsubramamian, C. V., Mashelkar, R. A., and Ulbrecht, J. J. (1978). Convective diffusion from a non-uniformly distributed source in non-Newtonian liquids, *Chem. Eng. Commun.*, **2**, 233.
43. Navari, R. M., Gainer, J. L., and Hall, K. R. (1971). Predictive theory for diffusion in polymer and protein solutions, *AIChE J.*, **17**, 1028.
44. Loucaides, R., and McManamey, W. J. (1973). Mass transfer into simulated fermentation media, *Chem. Eng. Sci.*, **28**, 2165.
45. Perez, J. F., and Sandall, O. C. (1973). Diffusivity measurement for gases in power-law non-Newtonian liquids, *AIChE J.*, **19**, 1073.
46. Konig, B., Lippert, J., and Schugerl, K. (1979). A steady state method for the determination of k_La in stirred tank reactors: oxygen transfer rate in model fermentation media with a high speed tube stirrer. *German Chemical Engineering*, **2**, 371.
47. Chapman, C. M., Gibilaro, L. G., and Nienow, A. W. (1982). A dynamic response technique for the estimation of gas–liquid mass transfer coefficients in a stirred vessel, *Chem. Eng. Sci.*, **37**, 891.
48. Chapman, C. M., Nienow, A. W., Cooke, M., and Middleton, J. C. (1983). "Particle–gas–liquid mixing in stirred vessels: part 4–mass transfer and final conclusions, *Chem. Eng. Res. Des.*, **61**, 182.
49. Heijnen, J. J., van't Riet, K., and Wolthius, A. J. (1980). Influence of very small bubbles on the dynamic k_La measurement in viscous gas–liquid systems, *Biotechnol. Bioeng.*, **22**, 1945.

50. Ranade, V. R., and Ulbrecht, J. J. (1977). Gas dispersion in agitated inelastic and viscoelastic liquids, *Proc. 2nd European Mixing Conf.*, BHRA Cranfield, Bedford, England, p. F6-83.
51. Hickman, A. D. (1985). Agitation, mixing and mass transfer in simulated high viscosity fermentation broths, Ph.D. Thesis, University of Birmingham, U.K.
52. Warmoeskerken, M. C. G., and Smith, J. M. (1983). Pitched blade turbine performance in gas–liquid systems, *Paper to Engineering Found. Conf. on Mixing*, Henniker, New Hampshire.
53. Zlokarnik, M. (1978). Sorption characteristics of gas–liquid contacting in mixing vessels, *Adv. Biochem. Eng.*, **8**, 244.

Chapter 7

Dispersion of Liquids in Liquids with Chemical Reaction

LAWRENCE L. TAVLARIDES

*Department of Chemical Engineering, Syracuse University, Syracuse,
NY 13210, USA*

1. INTRODUCTION

Liquid–liquid reaction systems form an important class of chemical
reactions which have significant importance in the chemical,
petroleum, mining, food, and pharmaceutical industries. Examples
of such systems are hydrometallurgical solvent extraction, phase
transfer catalysis, nitration–halogenation of hydrocarbons, emulsion
polymerization, hydrodesulfurization of crude stocks, hydrocarbon
fermentations and glycerolysis of fats. In order to develop a rational
design of chemical reactors for liquid–liquid systems, one must
describe the microscopic problem of interphase mass transfer with
reaction and couple this to the reactor design equations. The
microscopic problem is concerned with the determination of the
local rate of transfer of reactants and/or products between the two
phases as governed by the local bulk-phase conditions of concentration and temperature. This problem requires the solution of the
continuity equation for each species coupled with the local conservation of energy. The macroscopic problem is concerned with the
analysis of the evolution of both liquid phases through the volume
of the reactor, and makes use of the modelling of the local conditions obtained from a solution to the microscopic problem. This
problem requires use of the macroscopic material and energy

balances which should include an account of the interphase surface area, dispersed phase mixing, and dispersed phase holdup.

The scope of this work is to highlight the approaches taken to solve various aspects of the microscopic and macroscopic problems. For the former problem several important chemical reactions are discussed, various solutions of the microscopic problem for nonlinear systems are presented, and laboratory reactors used to determine kinetic models are examined. For the latter problem, noninteraction and interaction models will be presented. Use of the interaction or coalescence–redispersion models to describe the dispersed phase micromixing and concomitant effects on conversion will be demonstrated. References helpful in the structure of this paper are[1-5].

2. MICROSCOPIC PROBLEM

The differential equations of change must be written for each liquid phase. The conservation of species and energy equations for n-component systems become

$$\frac{\partial C_i^\alpha}{dt} + \mathbf{v} \cdot \nabla C_i^\alpha = D_i^\alpha \nabla^2 C_i^\alpha + R_i^\alpha(\mathbf{C}^\alpha, T^\alpha) \tag{1}$$

$$\frac{\partial T^\alpha}{\partial t} + \mathbf{v} \cdot \nabla T^\alpha = \epsilon^\alpha \nabla^2 T + \frac{\Delta H^\alpha}{\rho^\alpha C_p^\alpha} R_i^\alpha(\mathbf{C}^\alpha, T^\alpha) \tag{2}$$

where the assumptions of constant physical properties and no thermodynamic coupling is considered. The notation follows the text. We see for liquid–liquid systems a set of equations are needed for each phase.

When the bulk of both liquid phases are considered well mixed, integration of Eqs. (1) and (2) over the region of the interface gives expressions to calculate the interfacial flux. It is often assumed that there are no temperature gradients in this region so that the microscopic problem is reduced to the integration of the set of equations (1).

The problem is still formidable since (a) Eqs. (1) are nonlinear and coupled through the reaction term $R_i^\alpha(\mathbf{C}^\alpha, T^\alpha)$ and (b) understanding of the fluid mechanics is needed to account for the convective term $\mathbf{v} \cdot \nabla C_i$. Additional complications arise if the reaction occurs at the interface. Here the interphase boundary conditions may be nonlinear.

2.1. Chemical Reactions

The microscopic problem has generally been solved by assuming simple isothermal forms of the kinetic equations. Unfortunately, this is not true for real systems. The form of the intrinsic rate expressions $R_i^\alpha(C, T)$ must be determined over the concentration and temperature space to be expected between the liquid interface and the respective bulk phases. This problem requires generation of meaningful kinetic and thermodynamic equilibrium data with appropriate analysis of the experimental reactor. Much activity is under way in this area. It is appropriate to highlight some chemical systems and experimental methods employed to obtain the kinetic data.

2.1.1. Metal Chelation Reactions

A significant amount of attention has been given to these reactions because of their importance to hydrometallurgical processes[6-16]. Copper extraction from aqueous sulfuric acid solutions (obtained from leaching of ore) by organic chelation acids in an organic solvent is a prime example. The stoichiometry of the overall chelation reaction is

$$(Cu^{+2}) + 2\,\overline{HR} \rightleftharpoons \overline{CuR_2} + 2(H^+) \tag{3}$$

where the bar quantities refer to the organic phases, $\overline{HR}$ is the chelation agent, and $\overline{CuR_2}$ is complexed copper. Two chelation molecules of particular interest are those of β-hydroxy benzophenone (LIX 64B) and alkylated hydroxy quinoline as shown in Fig. 1. It is noted that the above is an equilibrium limited reaction which is pH sensitive and represents the overall distribution of copper between the phases.

Workers recognize the need to establish a suitable model for the chemical equilibrium[17-21]. For example, Bauer and Chapman[19] determined a chemically based model for the equilibrium constant in terms of total sulfuric acid and ionic strength. Thus

$$K_c = \frac{[\overline{CuR_2}][H_2SO_4]^2}{[CuSO_4][\overline{HR}]^2} = f(T, I, H_2SO_4, CuR_2) \tag{4}$$

Hoh and Bautista[20] obtained an equilibrium relationship for the

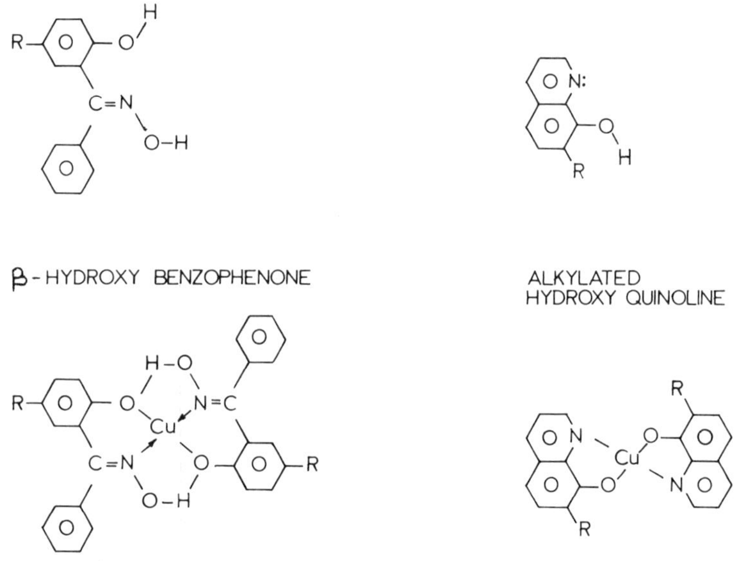

β-HYDROXY BENZOPHENONE

ALKYLATED
HYDROXY QUINOLINE

Figure 1   β-hydroxy benzophenone and alkylated hydroxy quinoline.

distribution of the metal between the phases in terms of K_c, the degree of formation of the metal ion, and an effective equilibrium constant. Whewell and Hughes[21] produced a similar expression which accounted for the actual H^+ ion concentration and polymerization of the chelation molecule in the organic phase. Lee and Tavlarides [22] and Agarwal and Tavlarides [23] give equilibrium models in terms of the concentration of extracted ion species which also consider aqueous phase equilibria.

Various kinetic expressions for these reactions have been proposed for the copper ion extraction from sulfuric acid solutions with 8-hydroxy quinoline[8-10], or β-hydroxy benzophenone [11], or the mixture of β-hydroxy benzophenone and an aliphatic oxime[8, 12-15]. It appears that the reaction occurs at the interface. Thus, Flett et al.[14] proposed a kinetic expression for the rate of copper extraction for the last system mentioned of the form

$$R_{(Cu^{++})} = -k_f \frac{[\overline{HR_{65}}][\overline{HR_{63}}]^{1/2}[Cu^{++}]}{[H^+]} + k_r \frac{[\overline{HR_{63}}]^{1/2}[H^+][\overline{Cu(R_{65})_2}]}{[\overline{HR_{65}}]}$$

(5)

The above expression is nonlinear and is based on a mechanism for a surface reaction. Concentrations must be known at the interface when diffusional resistances are important. The multicomponent rate expression couples Eqs. (1) for both phases through the interphase boundary conditions. Recently Freeman and Tavlarides[24] and Lee and Tavlarides[25] presented intrinsic kinetic rate models for the extraction of copper from sulfate solutions with a β-hydroxy oxime system and iron from sulfate solutions with a β-alkenyl-8-hydroxy quinoline system.

2.1.2. Nitration Reactions

The nitration of aromatic hydrocarbons by aqueous mixtures of nitric and sulfuric acids is an important industrial reaction[26-31]. Three successive regions of kinetic, slow reaction diffusional, and fast reaction diffusional were observed by Cox et al.[29-31] in the two-phase nitration of toluene as the sulfuric acid strength is increased. This reaction is an example of the reaction occurring in the film or bulk of the aqueous phase. Ample evidence indicates these reactions proceed via the reversible formation of the nitronium ion, NO_2^+,

$$HNO_3 + H^+ \underset{k_{-1}}{\overset{k_1}{\rightleftharpoons}} NO_2^+ + H_2O \qquad (6)$$

followed by attack of the NO_2^+ on the aromatic substrate

$$NO_2^+ + C_6H_5CH_3 \overset{k_2}{\rightarrow} NO_2C_6H_4CH_3 + H^+ \qquad (7)$$

The authors point out that in the range of concentrations they studied, diffusional resistances in the aqueous phase are important. In addition, as the concentration of sulfuric acid is increased there is a transition in kinetic order from first to zero with respect to toluene. Thus Eq. (1) must be solved with $R_{NT}^a = kC_N C_T^n$ with $n = 0$ or 1 for the aqueous phase when the organic phase is pure toluene.

2.1.3. Phase Transfer Catalytic Reactions

Phase transfer catalysis[32-34] is a recently developed organic synthesis technique which accelerates the reaction rate for a heterogeneous reaction between two reactants in separate liquid phases which are sparingly soluble in opposite phases. The reaction normally occurs at the interface between the phases and is extremely slow. The equations below depict the process:

Aqueous phase

$$[Q^+X^-] + A^- \underset{k_{-1}}{\overset{k_1}{\rightleftharpoons}} [Q^+A^-] + X^-$$

Phase Boundary

$$- -\mathcal{J}\!\!\mathfrak{l} - - - -\!\!\!\mathfrak{l}\!\!\mathfrak{l} - - \tag{8}$$

Organic phase

$$[Q^+X^-] + AB \underset{k_{-2}}{\overset{k_2}{\rightleftharpoons}} [Q^+A^-] + BX$$

Thus, in the reaction involving the water soluble nucleophile A^-, the addition of the phase transfer catalyst, Q^+X^-, causes the transfer of the nucleophile as an ion-pair $[Q^+A^-]$ into the organic phase where it reacts with the organic reagent, BX. Migration of the cationic catalyst back to the aqueous phase completes the cycle. The success of the catalytic effect depends to a large extent upon the high partition coefficient of the ion pair $[Q^+A^-]$ compared with that of $[Q^+X^-]$ between the two phases. Typical catalysts are quarternary ammonium or phosphonium salts.

It is observed that reversible reactions occur in both phases limited by chemical equilibria. Furthermore, interphase transfer of ion-pairs in both directions occurs. The reactions may occur in the bulk phase or in a film region near the interphase when diffusional resistances become important (high catalyst concentrations). Thus it may be necessary to apply Eqs. (1) to both phases.

2.2. Fluid Mechanics

The simplest models of the fluid mechanics of a liquid phase in the vicinity of a liquid–liquid interface are based on the assumption that the velocity vector v^α is parallel to the interface and time independent. The concentration gradient vector ∇C_i^α is, in addition, assumed orthogonal to the interface. Thus, the convective terms drop out of Eqs. (1) and give rise to the classical film theory and penetration theory models. The fluid mechanics do not drop out even in these simple models, since some adjustable parameter (the film depth or the penetration time or distance) must be considered based on the prevailing flow pattern. The values of these parameters can, as a first approximation, be obtained from experimental evidence on simple mass transfer without chemical reaction.

The use of the penetration model applied to second-order irreversible reaction (similar to nitrations) was solved by van Krevelen and Hoftijzer[35] to obtain expressions for the enhancement of interphase mass transfer due to reaction. For example, assuming that reaction occurs only in the aqueous phase, the organic phase is pure reactant, and isothermal conditions Eqs. (1) become (superscripts are dropped for convenience)

$$\frac{\partial C_A}{\partial t} = D_A \frac{\partial^2 C_A}{\partial x^2} - k_2 C_A C_B \tag{9}$$

$$\frac{\partial C_B}{\partial t} = D_B \frac{\partial^2 C_B}{\partial x^2} - z k_2 C_A C_B \tag{10}$$

subjected to

$$t = 0, x > 0: \quad C_A = 0; \ C_B = C_{B,0}$$

$$t \geq 0, x = 0: \quad C_A = C_{A^*}; \ \frac{\partial C_B}{\partial x} = 0 \tag{11}$$

$$t \geq 0, x = \infty: \quad C_A = 0; \ C_B = C_{B,0}$$

The solution is presented graphically in Fig. 2 in a condensed form. Here the enhancement factors E for second order reaction and E_i for instantaneous reaction are given by

$$E = \frac{N_A(\text{with second-order reaction})}{N_A(\text{physical diffusion})}$$

$$= \sqrt{\left(M' \frac{E_i - E}{E_i - 1} \right)} \bigg/ \tanh \sqrt{\left(M' \frac{E_i - E}{E_i - 1} \right)} \tag{12}$$

$$E_i = f\left(\frac{C_{B,0}}{z C_{A^*}} \sqrt{\frac{D_B}{D_A}} \right) \tag{13}$$

and

$$M' = \pi/4 k_2 C_{B,0} t_c \tag{14}$$

From these solutions, the local rate of enhanced mass transfer to a phase can be calculated as a function of the local bulk phase concentrations, physical properties, fluid dynamic parameters, and chemical kinetics. It should be noted that the apparent local rate of production of a reactant $\overline{R}_A$ can be calculated from the interfacial flux

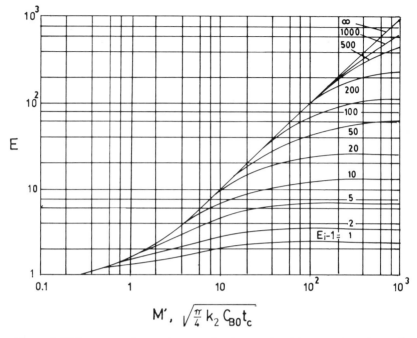

$$M', \sqrt{\tfrac{\pi}{4} k_2 C_{B0} t_c}$$

Figure 2   Enhancement factors for second order reaction (after van Krevelen and Hoftijzer[25]).

$$\overline{R}_A = N_A a^* \qquad (15)$$

where N_A is the interfacial molar flux of A and a^* is the interfacial surface area per volume of dispersion.

The important case of simultaneous diffusion and chemical reaction in a liquid–liquid system when these processes are occurring in both phases has been solved by approximate analytical methods[36–39]. These workers used the film theory to obtain expressions to estimate the total rate of reaction occurring in both phases and the approximations for the enhancement factors. Mhaskar and Sharma[37] classified the various reaction–diffusion controlled situations which exist by considering the relative values of the rates of diffusion and chemical reaction. These are subdivided into reactions occurring only in the bulk or reactions occurring in the films adjacent to the interface with simultaneous diffusion of reacting species. Merchuk and Farina[38] and Sada et al.[39] present approximate solution of the

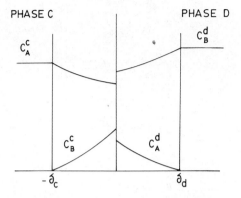

Figure 3   Idealized representation of film model for components A and B.

case where diffusion and second-order irreversible reaction are important in both phases. The situation is shown diagrammatically in Fig. 3. The equations solved were

Phase c: $-\delta_c < x < 0$

$$D_A^c \frac{d^2 C_A^c}{dx^2} = k_c C_A^c C_B^c \qquad (16)$$

$$D_B^c \frac{d^2 C_B^c}{dx^2} = k_c C_A^c C_B^c \qquad (17)$$

Phase d: $0 < x < \delta_d$

$$D_A^d \frac{d^2 C_A^d}{dx^2} = k_d C_A^d C_B^d \qquad (18)$$

$$D_B^d \frac{d^2 C_B^d}{dx^2} = k_d C_A^d C_B^d \qquad (19)$$

Subjected to

$$x = -\delta_c \quad C_A^c = C_{A,0}^c$$

$$C_B^c = 0 \qquad (20)$$

$$x = 0 \qquad D_A^c \frac{dC_A^c}{dx}\bigg|_{x=0} = D_A^d \frac{dC_A^d}{dx}\bigg|_{x=0} \qquad (21)$$

$$D_B^c \frac{dC_B^c}{dx}\bigg|_{x=0} = D_B^d \frac{dC_B^d}{dx}\bigg|_{x=0}$$

$$C_A^d = h_A C_A^c$$

$$C_B^c = h_B C_B^d$$

$$x = \delta_d \qquad C_A^d = 0 \qquad (22)$$

$$C_B^d = C_{B,0}^d$$

Values of the enhancement factors were calculated and were compared to values obtained with an exact numerical solution. The analytical approximation technique of Merchuk and Farina involves linearization of the reaction terms with values of concentrations of one of the reactants at the interface. The technique of Sada et al. is based on assuming concentration profiles based on the diffusion equation without reaction. Over the range of parameters studied, values of the enhancement factors of A and B calculated by the approximate analytical method agreed to within 6% of values from numerical solutions in the work of Sada et al. Figure 4 is a plot of the enhancement factors. Parameter definitions are given in Table 1. The benefit of the use of the approximate analytical methods is that they can be employed with the design equations of the macroscopic problem directly.

Studies on interface mass transfer with reaction occurring at the interface is receiving much attention recently due to the application to hydrometallurgical processes[40–43]. It is noted that three general cases exist depending upon the relative rates of the mass transfer to reaction processes: (a) slow reaction where diffusional resistances are unimportant, (b) intermediate reaction at the interface where the flux of a diffusing species to the interface equals the rate of surface reaction and (c) instantaneous surface reaction where the

Table 1   Parameters for Fig. 4, from Sada et al.[39]

$q_d = \nu_d C_{A,0}^c / C_{B,0}^d$	$Z_A = k_{LA,c}/k_{LA,d}$
$q_c = \nu_c C_{A,0}^c / C_{B,0}^d$	$Z_B = k_{LB,d}/k_{LB,c}$
$r_d = D_A^d/D_B^d$	$M_c = k_c C_{B,0}^d \delta_c^2 / D_A^c$
$r_c = D_A^c/D_B^d$	$M_d = k_d C_{B,0}^d \delta_d^2 / D_A^d$

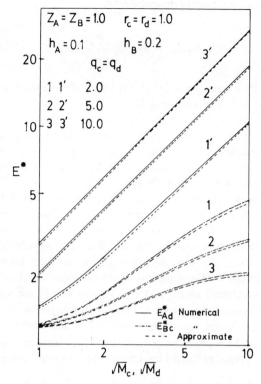

Figure 4 Comparison of enhancement factors for second order reactions in both phases of approximate analytical solution with numerical solution (after Sada et al.[39]).

condition of chemical equilibrium must be satisfied at the interface.

England and Berg[40] and Yagodin et al.[41] present analytical solutions for a penetration theory model. The transport problem is diffusion of reactant A from phase c to the interphase, irreversible or reversible first-order surface reaction (or adsorption kinetics) through an intermediate B which exists only at the interface, and diffusion of the product C (or desorbed species) into phase d. The diffusing species were assumed to be insoluble in the opposite phases. The interfacial boundary condition for equality of fluxes in this problem has an accumulation term, viz.

$$D_A^c \frac{\partial C_A^c}{\partial x} = -D_C^d \frac{\partial C_c^d}{\partial x} + \frac{d\Gamma}{dt}, \quad x = 0 \tag{23}$$

where Γ = surface concentration of the intermediate.

The surface reaction is described by

$$A_{\text{phase }c} \underset{k_1}{\overset{k_1}{\rightleftharpoons}} B_{\text{surface}} \underset{k_{-2}}{\overset{k_2}{\rightleftharpoons}} C_{\text{phase }d} \tag{24}$$

which permits expressions for $d\Gamma/dt$ to be determined assuming the adsorbed layer has a thickness δ^*.

Chapman et al.[42, 43] employed the two-film theory and the penetration model for metal extraction with instantaneous reversible interfacial reaction as described by Eqs. (3) and (4). The process is depected in Fig. 5. The unique feature of metal extraction mass transfer is the way in which the multicomponent nature of the system strongly couples the concentrations and fluxes through the equilibrium and the stoichiometry of the exchange reaction. It is necessary to consider the mass transfer of all solute species simultaneously rather than treating the metal species as solutes in pseudo-binary solutions. As the diffusion equations are uncoupled, the problem reduces to the solution of a nonlinear algebraic equation. Values of interfacial metal flux can be calculated for various bulk-phase conditions. These values can be employed with the macroscopic reactor design equations to predict reactor performance.

To account for spherical geometry of the droplet and the effects of dispersed phase holdup Gas-Or and Hoelscher[44] solved the prob-

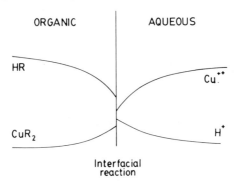

$$Cu^{\bullet\bullet} + \overline{HR} \rightleftharpoons \overline{CuR_2} + 2H^+$$

$$K_{c,eq} = \frac{[CuR_2]_i [H^+]_i^2}{[Cu^{\bullet\bullet}]_i [HR]_i}$$

Figure 5   Concentration profiles during metal extraction.

lem of unsteady molecular diffusion from a drop with first order irreversible reaction in a finite fluid element employing the spherical cell model. The spherical cell model assumes that in a turbulent dispersion an average drop can be envisioned about which there exists an equipotential surface through which no mass transfer occurs. This equipotential surface is assumed to be spherical. Solution of the appropriate conservation equation yields expressions for the interfacial flux in terms of the drop radius. Tavlarides and Gal-Or[45, 46] extended these concepts to multicomponent mass transfer with general reversible first order homogeneous reactions occurring in both phases. Along this line, Johnson and Akehata[47] consider second-order reaction accompanied by diffusion from a stationary sphere in a stagnant fluid. An approximate solution to the steady-state problem was obtained and compared with previous values of the enhancement factors.

It is necessary to mention various attempts to account for the convective term in Eqs. (1) before concluding approaches to solve the microscopic problem. Noteworthy is the work by Ruckenstein et al.[48] which solves two cases of the problem of unsteady convective-diffusion with irreversible homogeneous first-order chemical reaction in either phase. Analytical solutions of the average rate of mass transfer over the surface of the drop are presented and shown to compare well with the numerical solutions. In addition Brunson and Wellek[49] developed a numerical solution for mass transfer inside liquid droplet accompanied by second-order chemical reaction. Charts of total rate of mass transfer vs. dimensionless time were prepared from the solution of the unsteady-convective diffusion equation.

3. LABORATORY REACTORS FOR KINETIC STUDIES

There are many devices used to study two-phase mass transfer with reaction at or near a liquid–liquid interface. The properties a liquid–liquid contactor must have for successful kinetic studies are: a well defined surface area, a clean freshly formed interface, well defined hydrodynamics without significant and effects, and a short contact time.

Some of the more prevalent techniques used and their disadvantages are shown in Table 2. These techniques included a dispersed phase contactor, Lewis-cell where two fluid layers are in-

Table 2   Experimental Contactors for Liquid–Liquid Kinetic Data

Contactor	Description	Comments	References
AKUFVE	Stirred cell	Surface area and mixing ill defined, hydrodynamics unknown	8, 9, 16
Lewis-cell	Fluids layered in cell, each layer stirred separately	Accumulation of surfactants, surface waves complicate hydrodynamics, data scatter of ± 30%	52–54
Diaphragm cell	Fluids partitioned by diaphragm	Surface area unknown in porous membrane	
Single drop experiment	Drop falls (rises) in liquid	Three hydrodynamic regimes, to complicated for simplified analysis, surfactant contamination	11–13, 15
Growing drop	Drop formation in liquid	Complex hydrodynamics, surface contamination	10
Liquid jet recycle reactor	Liquid jet with cocurrent flow of second phase	Surface area defined, easy cleaning of interface, hydrodynamics accountable, instantaneous monitor of circulating fluid.	50, 51

dividually stirred, diaphragm cell, single drop experiment, growing drop, and laminar liquid jet.

Of these techniques the newly developed liquid jet recycle reactor (LJRR)[24, 25, 50, 51], appears to offer significant advantages. In this technique a liquid jet is suspended between a capillary nozzle and capillary receiver. The outer fluid is passed co-currently to the jet in a reactant chamber and recycled continuously throughout the course of the experiment. Constant UV-visible monitoring of the reaction fluids permits instantaneous recording of changes in the organic species concentrations. Figure 6 is a diagrammatic representation of the apparatus. Figure 7 is a photograph of the jet during operation to obtain rate data for the hydrometallurgical surface reaction of copper ions in sulfuric acid solutions with β-hydroxy benzophenone in toluene. The advantages of this technique include:

1. Marangoni type instabilities can be reduced or eliminated due to short contact times (range of 0.05 to 0.5 sec.).
2. Interfacial impurities can be removed without disturbing the interface.

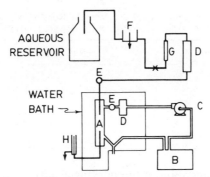

Figure 6   Liquid jet recycle reactor apparatus. (A) jet chamber, (B) recording spectrophotometer, with flow cell, (C) pump, (D) heat exchanger, (E) Thermometer well, (F) constant head vessel, (G) rotometer, (H) leveling device.

Figure 7   Photograph of liquid–liquid jet, jet flow rate = $90°$ ml-min^{-1}, outer fluid flow rate = 28.9 ml-min^{1}.

Table 3   Reproducibility of Liquid Jet Recycle Reactor, Constant Liquid Compositions, [LIX 65] = 0.567, [CuSO$_4$] = 0.125, [H$_2$SO$_4$] = 0.016. (molar).

Run No.	Rate, g mole-cm^{-2}-sec^{-1} × 10^8
1	1.50
2	1.41
3	1.45
4	1.41
Average	1.44
Std. Dev.	0.04

3. Precise determination of the interfacial surface area.
4. Easy phase separation (end effects are not significant).
5. Geometry permits analysis of the hydrodynamics.
6. Readily adapted to adsorption–desorption studies.

Reproducibility of data and inherent error is extremely good for liquid–liquid systems. For example, Table 3 shows reaction rates reproducible within 3% for one standard deviation about the mean value.

The LJRR can be employed to measure reactions whose rates are of the order of the diffusion process or slower. Interfacial reactions with half lives as short as 0.035 to 0.06 seconds would be modelled as instantaneous. The lowest rate is restricted by the jet stability. The jet can be sustained up to four hours with little difficulty for which the lowest reaction rate to study would be about 1.0×10^{-8} moles/cm^2 sec.

4.   MACROSCOPIC PROBLEM

The macroscopic problem deals with the reactor design question. A knowledge of the mixing conditions in both phases is of prime importance. Two levels of approach to the problem emerge. *Non-interaction* models are employed when the micromixing of either phase is unimportant. *Interaction of coalescence and dispersion* models are applied to problems where micromixing (particularly the dispersed phase) becomes important.

4.1.   Noninteraction Models

The generalized multiphase reactor model of Pavlica and Olson[57] can be applied to tubular reactors such as motionless mixers or column extractors. For the case of constant volumetric flow of both phases, isothermal conditions and steady-state operation one conservation of species equation is needed for each component in each phase. If axial but no radial dispersion is present, then the following equations result:

Dispersed phase:

$$\frac{1}{Pe_d} \frac{d^2 C_i^d}{dz^2} = \frac{dC_i^d}{dz} - \frac{\Phi L}{u_d} R_i^d - \frac{K_{d,i} a^* L}{u_d} (\bar{D} C_i^d - C_i^c) \qquad (25)$$

Continuous phase:

$$\frac{1}{Pe_c} \frac{d^2 C_i^c}{dz^2} = \pm \frac{dC_i^c}{dz} - \frac{(1-\Phi)L}{u_c} R_i^c + \frac{K_{d,i} a^* L}{u_c} (\bar{D} C_i^d - C_i^c) \qquad (26)$$

where Φ = dispersed phase holdup; a^* = interfacial surface area per unit volume of dispersion; L = reactor length; $\bar{D}$ = equilibrium distribution coefficient; and $K_{d,i}$ = overall mass transfer coefficient of species i based on the dispersed phase, and the other terms are in the notation. In Eq. (26) the first term on the right hand side is negative for countercurrent flow and positive for cocurrent flow.

This flow model description assumes complete radial mixing in both phases. In particular the dispersed phase droplets are uniform in concentration regardless of size at any point radially in the reactor. It is noted that in this formulation it is assumed that the reaction is slow in both phases and that mass transfer is in series with bulk-phase reaction. For the general case of two-phase diffusion with reaction, the reaction term is incorporated into the mass transfer term by correcting the mass transfer coefficient with the appropriate enhancement factor. For these cases solutions to the microscopic problem as discussed earlier must be applied.

Similar equations can be written for the CSTR where both phases are perfectly mixed. The unsteady problem under isothermal conditions can be posed as

Dispersed phase:

$$\frac{dC_i^d}{dt} = \frac{(C_{i0}^d - C_i^d)}{\Phi \tau^d} + R_i^d - \frac{K_{k,i} a^*}{\Phi} (\bar{D} C_i^d - C_i^c) \qquad (27)$$

Continuous phase:

$$\frac{dC_i^c}{dt} = \frac{(C_{i0}^c - C_i^c)}{(1-\Phi)\tau^c} + R_i^c \frac{K_{d,i} a^*}{\Phi} (\bar{D} C_i^d - C_i^c) \qquad (28)$$

where $C_{i,0}^{\alpha}$ is the inlet concentration of species i in phase α are τ^{α} is the average residence time of phase α. Both phases are considered perfectly mixed on the microscale level. Again it has been assumed that the mass transfer is in series with bulk-phase reaction.

Note that Eqs. (25)–(26) and (27)–(28) contain the interfacial surface area, a^*, which is related to the volume to surface average drop diameter d_{32} by

$$a^* = \frac{6\Phi}{d_{32}} \qquad (29)$$

In order to properly apply these equations, information on the average drop size must be employed. Expressions for d_{32} as a function of physical parameters and operation conditions have been obtained by numerous investigations whose efforts were summarized by Tavlarides and Stamatoudis[5]. The most frequently reported correlation for a stirred tank is of the form

$$\frac{d_{32}}{D_I} = a_1(1 + a_2\Phi)N_{We}^{-a_3} \qquad (30)$$

where D_I is the impeller diameter, N_{We} is the Weber number, and the a_i's are constants.

Solutions of Eqs. (25)–(26) and (27)–(28) subjected to the appropriate subsidiary conditions will yield expected conversions and selectivity for appropriate values of the design equation parameters. The dynamic behavior of an agitated two-phase reactor with nonisothermal reaction and dynamic variations in drop diameter recently studied by Barnea, Hoffer and Resnick[58] is an example of the use of Eqs. (27)–(28) to predict dynamic reactor performance. Another study by Sepulveda and Miller[59] applied these equations to multiple metal ion extraction in continuous flow mixers. Here a set of these equations were solved for each stirred reactor in a series of tanks. Results permitted the estimation of operation parameters for optimum selectivity of the desired metal ion.

These models are the first level for dispersed phase reactors. They permit calculations of conversion and selectivity for column and stirred vessel extractors. However, they are most applicable for slow reactions when fluid mixing effects are less significant. They do not account for dispersed or continuous phase mixing and the effects of drop size distributions on mass transfer rates and interfacial flux.

A second level of noninteraction models includes the use of drop size distributions and residence time distributions of the droplets in the vessel. Notable contributions are those by Gal-Or and co-workers[44, 45, 60] and similar approaches by Korchinsky and co-workers[61, 62]. In the former models the spherical cell concepts mentioned earlier are employed. Expressions for interfacial flux are developed based on the appropriate microscopic model. The average moles transferred per unit volume of dispersion is obtained by integration of the flux over the distribution of droplet sizes for an average residence time of the drops in the reactor. The residence time distribution of droplets could also be accounted for by an "integral transformation" technique. These models can account for the drop size distributions and, in part, the effects of holdup on mass transfer rate and interfacial flux. These models represent generally the fluid mixing case of completely mixed continuous phase and segregated dispersed phase.

Korchinski and co-workers modelled mass transfer in extraction columns by assuming axial dispersion characterizes the continuous phase behavior, and plug flow with "forward mixing", that is, with residence time varying with drop size, characterizes the dispersed phase. The model equations for diffusion of a single species are

Dispersed phase

$$\frac{dy_i}{dz} = \frac{6K_{OD,i}L}{u_{d,i}d_i}(mw - y_i) \quad i = 1 \text{ to } 1 \tag{31}$$

Continuous phase

$$\frac{1}{Pe_c}\frac{d^2w}{dz^2} = -\frac{dw}{dz} + \frac{6\Phi L\rho_d}{u_c\rho_c}\sum\frac{K_{OD,i}f_i(mw - y_i)}{d_i} \tag{32}$$

with the boundary conditions

$$z = 0, \ w = w_1, \frac{dw}{dz} = 0, \ y_i = \bar{y}_1 \tag{33}$$

$$z \doteq 1, \ w = w_2 - \frac{1}{Pe_c}\frac{dw}{dz}\bigg|_{z=1}$$

$$y_j = \bar{y}_2 = \sum\frac{f_i u_{d,i} y_i \Phi}{u_d} \tag{34}$$

Here w and y are the weight fractions of the solute in the dispersed and continuous phases. To utilize these equations, correlations of data for the continuous-phase axial dispersion coefficient and the overall mass transfer coefficient based on the two film model, $K_{OD,i}(d_i)$ are needed, in addition to the size distribution, $f_i(d_i)$, and the hydrodynamic parameters to specify $u_{d,i}$. While these models were applied to the case of mass transfer, the coefficient $K_{DO,i}$ can be determined for reaction cases as mentioned in the microscopic problem. The model does account for drop size distribution on mass transfer rate and the drop size on the interfacial flux. Further, an attempt to account for the droplet residence time is made through the use of hydrodynamic expressions for the velocity of droplets of varying size. Results of extraction efficiencies versus column heights were obtained. These models represent a significant step towards accounting for the dispersed phase hydrodynamics. However, the micromixing phenomenon between droplets due to coalescence were not properly included. Further, the relation of the parameter models to describe droplet size distributions and hydrodynamics to the operation conditions were not prescribed.

4.2.   Interaction Models

Interaction models are the most powerful for modelling reactions in liquid dispersions and can play an important role in predicting conversion and selectivity when dispersed phase mixing is important on the transport processes. Further, the framework of these models permits a rational basis to account for the macromixing phenomena. A detailed review of these models is to appear elsewhere[5]. Three types of interaction models are population balance models, Monte Carlo coalescence–dispersion (C–D) simulation techniques, and a combination of macromixing and micromixing concepts with Monte Carlo simulations.

The population balance approach has found great utility in description of the droplet dynamics in various flow fields for agitated vessels[63–71] and for predicting droplet mixing on conversion and interphase mass transfer[65, 72–79].

Hulbert and Katz[79] developed a framework for the analysis of particulate systems with the population balance equation (PBE) for a multivariate particle number density. This number density is defined over phase space which is characterized by a vector of the least number of independent coordinates attached to the particle distribution which allow complete description of the properties of the distribution. Phase space is composed of three external particle coordinates (x_{e1}, x_{e2}, x_{e3}), which refer to the spatial distribution of particles, and m internal particle coordinates, $(x_{i1}, x_{i2}, \ldots, x_{im})$, which permit a quantitative description of the state of an individual particle such as its mass, concentration, temperature, etc. In the case of a homogeneous dispersion, such as a well mixed vessel, the external coordinates are unnecessary, whereas, for a nonideal stirred vessel or tubular configuration they are required to describe the spatial variation in the particle number density. Let $NA(\mathbf{x}; t)d\mathbf{x}$ represent the number of particles per unit volume of dispersion at time t in the incremental range $\mathbf{x}$ to $\mathbf{x} + d\mathbf{x}$. Then the number density continuity equation in particle phase space is

$$\frac{\partial}{\partial t} NA(\mathbf{x}; t) + \nabla \cdot [\dot{\mathbf{x}}NA(\mathbf{x}; t)] = h^+(\mathbf{x}; t) - h^-(\mathbf{x}; t) \qquad (35)$$

where $\dot{\mathbf{x}}$ is the coordinate velocity in phase space, and $h^+(\mathbf{x}; t)\, d\mathbf{x}$ and $h^-(\mathbf{x}; t)\, d\mathbf{x}$ are the number of droplets which are produced and destroyed, respectively, per unit volume of dispersion at time t in the incremental range $\mathbf{x}$ to $\mathbf{x} + d\mathbf{x}$. Solution of Eq. (35) requires specification of the coordinate velocities through the mass, energy, and momentum balances.

Coulaloglou and Tavlarides[68] employed Eq. (35) to test phenomenological models to describe drop breakup and coalescence in a turbulently agitated liquid–liquid dispersion. Experiments were conducted under isothermal conditions, physical equilibrated dispersions, uniform holdup, spatially homogeneous dispersions at steady-state operation for which Eq. (35) becomes

$$\int_v^{v_{\max}} \beta(v', v)\nu(v')g(v')N(t)A(v', t)\, dv' - g(v)N(t)A(v, t)$$

$$+ \int_0^{v/2} \lambda(v - v', v')h(v - v', v')N(t)A(v - v', t)N(t)A(v')\, dv'$$

$$- N(t)A(v, t) \int_0^{v_{\max}-v} \lambda(v, v')h(v, v')N(t)A(v', t)\, dv'$$

$$= N(t)A(v, t)f(v) - n_0(t)A_0(v, t). \qquad (36)$$

The first two terms on the left-hand side represent the rate of formation and loss of droplets of size v, respectively, due to breakage. Here $g(v)$ is the breakage frequency, $\nu(v)$ represents the number of daughter droplets formed from breakage of size v drop, and $\beta(v', v)$ is the distribution of daughter droplets formed from breakage of size v' drop. The second and third terms represent the rate of formation and loss of droplets of size v due to coalescence. Here $\lambda(v, v')$ is the collision efficiency between drops of size v and v', and $h(v, v')$ is the collision frequency between drops of size v and v'.

The last two terms account for the flow of droplets of size v from and into the vessel where $f(v)$ is the escape frequency of droplet size v. Phenomenological models were proposed to describe the aforementioned breakage and coalescence rate functions. Comparison of results from Eq. (36) with experimental drop size distribution data and coalescence frequencies permitted a check of the validity of the models for the droplet rate processes. Figure 8 compares experimental and theoretical drop-size distributions and Fig. 9 compares coalescence frequencies. It should be noted that this approach permits interfacial surface area and droplet mixing frequencies to be predicted from physical parameters and operation conditions of the stirred vessel.

Curl[72] proposed a simplified homogeneous dispersed phase mixing model to study effects of dispersed phase mixing on a reaction occurring in the dispersed phase. Uniform drops are assumed, coalescence occurs at random and redispersion occurs immediately to yield equal size drops of the same concentration, and the dispersion is assumed to be homogeneous. Irreversible reaction of general order s was assumed to occur in the drops. The PBE given below is a subcase of Eq. (35) when the single internal coordinate is drop concentration such that $n(c)dc$ is the fraction of drops of concentration range c to $c + dc$. Thus.

$$\frac{1}{\omega_r} \frac{\partial n(c)}{\partial t} = n_0(c) - n(c) + I^* \left[4 \int_0^c n(c + \alpha)n(c - \alpha)\, d\alpha - n(c) \right]$$

$$+ K_r \frac{\partial [csn(c)]}{\partial c} \tag{37}$$

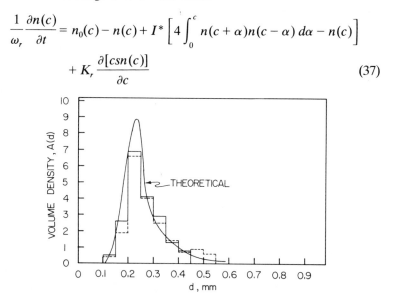

Figure 8 Theoretical and experimental size distributions—histogram, circulation region; N^*-310 rpm; $\Phi = 0.10$, $d_{32} = 0.255$ (exper.) vs. 0.250 (calc.), $\omega_v = 0.980$ (exper.) vs. 1.026 (calc.) (after Coulaloglou and Tavlarides[63]).

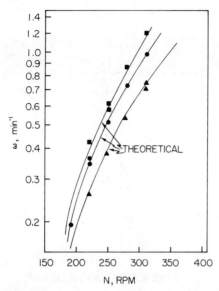

Figure 9   Theoretical and experimental coalescence frequency vs. impeller speed.
Solid lines are calculated. Symbols are experimental data for which $\Delta - \phi = 0.05$,
$\bullet - \phi = 0.10$, and $\blacksquare - \phi = 0.15$ (after Coulaloglou and Tavlarides[68]).

where $K_r = k/\omega_r c_0 =$ reaction modulus, $\omega_r =$ residence frequency,
$I^* = \omega_i/\omega_r$ dispersed phase mixing modulus, and $\omega_i =$ dispersed phase
mixing frequency.

Figure 10 shows the effect of dispersed phase mixing on conversion
for a zero-order reaction occurring in the drops (approximates
diffusion controlled reactions). It is seen that at 80% conversion,
mixing of the dispersed phase reduces the reactor volume by nearly
a factor of three. Similar results were shown by Shain[73] for second-
order reactions, and an analysis was made of selectivity for two
competing parallel reactions which occur in the dispersed phase.

It can be concluded that the population balance equations have
been most useful in delineating the role of droplet size distributions
and coalescence and breakage phenomena on mass transfer with
reactions. However, as the PBE models increase in dimensionality,
the solutions for multivariate density functions become formidable.

Statistical Monte Carlo coalescence–dispersion models were
developed to bypass difficulties in solution of the multivariate PBE.
It should be noted that simulation methods are computer al-
gorithms. These statistical models attempt to account for the
macromixing and micromixing phenomena. Spielman and Levenspiel[80] first applied such a model to study effects of coalescence and

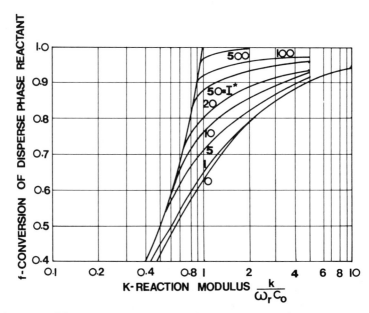

Figure 10   Effect of dispersed phase mixing on the conversion for a zero-order reaction in the drops in a single-stage two-phase, stirred reactor (after Curl[72]).

dispersion on the progress on zero and second-order reactions which occur in the dispersed phase. Other approaches have followed[81–92] and C–D modelling has also been applied to single-phase reactors.

The model by Zeitlin and Tavlarides[85–88], attempts to account for the macro-flow patterns of the dispersed phase in a turbulently agitated flow vessel, the droplet mixing phenomena on transport processes via breakage and coalescence, and nonuniform drop size on mass transfer or reaction in dispersions. The model is heterogeneous in nature and can account for spatial variations in dispersion properties and transport rates. While this conceptual model has been improved[89, 90, 93–95], it will be discussed here. Batch, semibatch or continuous flow can be simulated. The continuous phase is well mixed. Particle movement is either random or follows the flow direction of the sum of the local average fluid velocity and the particle gross terminal velocity. These actions are depicted in Figs. 11 and 12. The probability of droplet breakup is assigned based on droplet size. Binary breakage yields two random sized particles whose mass equals the parent drop. The probability of coalescence exists when two drops enter the same grid location. Particles are added and removed to simulate flow. The drops are segregated

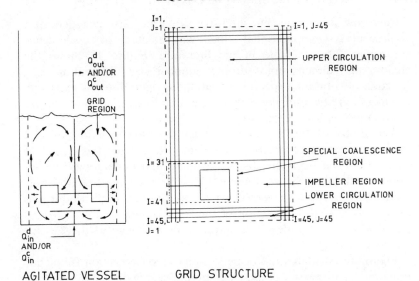

Figure 11 Pictorial representation of agitated vessel and grid (after Zeitlin and Tavlarides[85]).

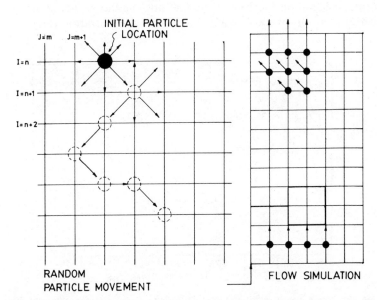

Figure 12 Representation of particle movements (after Zeitlin and Tavlarides[85]).

during flow simulation, but intensive properties are averaged after coalescence. Computational time was related to coalescence frequency data available in the literature. Figure 13 shows the steady state dimensionless droplet number size distribution as a function of rotational speed for continuous operation. As expected, the model predicts smaller droplet sizes and less variation of the size distribution with increase in rotational speed.

The utility of the model to predict the effects of interdroplet mixing on extent of reaction was demonstrated for the case of a solute diffusing from the dispersed phase and undergoing second-order reaction in the continuous phase. For comparison the normalized volumetric dispersed phase concentration distribution is defined as $f_v(\overset{*}{y})\,dy^*$ = fraction of the total volume of the dispersed phase with dimensionless concentration in the range y^* to $y^* + dy^*$ where $y^* = c/c_0$.

Figure 14 compares the dispersed-phase concentration frequency function for dispersed phase of complete segregation, intermediate mixing, and pseudo-completely mixed. Multiple peaks for the frequency function exist for segregated systems due to slow depletion of extractant from large drops in the system. The model permits the incorporation of both size distribution and droplet mixing effects on calculations of conversions. Further the simulation permits calculation of reaction time and agitation necessary to insure that all drops have reached a certain concentration range.

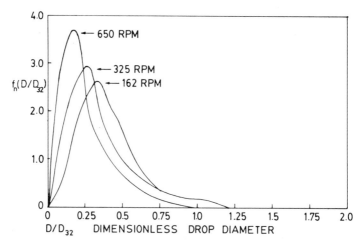

Figure 13   Number size distribution as a function of rpm (after Zeitlin and Tavlarides[85]).

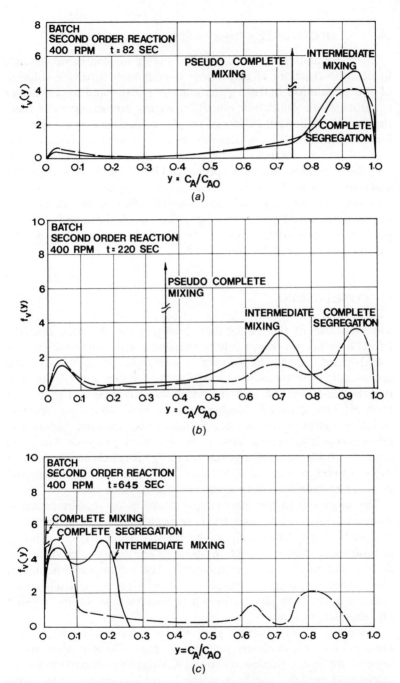

Figure 14 Volumetric dispersed phase concentration distribution for second-order reaction in continuous phase $k_r = 3.22 \times 10^2$ liter/g-mol, sec; $k_1 = 0.05$ cm/sec, $\Phi = 0.0063$. (a) – $t = 82$ sec, (b) $t = 220$ sec., (c) $t = 645$ sec.

265

These simulation methods appear to offer substantial advantages to describe reactions in dispersions over models involving solution of the population balance equations or macro- and micro-mixing concepts. With simulation models, complex multicomponent reactions can be followed. Effects of droplet coalescence and dispersion for nonuniform drops on transport processes can be accounted for, and effects of unmixed feeds on conversion and selectivity for complex reactions present no problem. Use of the models to predict scale-up of stirred vessels is not possible until the macro flow patterns and the local turbulence energy dissipation can be estimated with vessel geometry, operation conditions, and physical properties. Local coalescence and breakage frequencies can then be calculated from available models[67, 68].

5.   CONCLUSIONS

The objective of this paper is to highlight the direction research efforts have taken towards solving reaction engineering problems for liquid–liquid dispersions. To predict interphase transport rates, the microscopic problem must be solved with adequate understanding of the chemical reaction and the local fluid dynamics. The former problem has spawned development of new laboratory reactors which permit unambiguous kinetic data to be obtained. Substantial effort towards obtaining kinetic models is in progress. The fluid dynamic problems, insofar as reactor design is concerned, have been bypassed temporarily through use of established models of film and penetration theories. More work needs to be done here.

The approach to the macroscopic problem on reactor design is shown to consist of various levels of sophistication. The nature of the model used depends upon the relative values of the rates of diffusion and chemical reaction and thus the importance of micromixing and macromixing phenomena. The noninteraction models for a tubular reactor are based on knowledge of axial dispersion coefficients for both phases. Such data is lacking for various column and chemical systems.

Modelling of micromixing phenomena is limited to either complete mixing or completely segregated cases. These models are in general applicable to slow reactions. Coalescence–dispersion (C–D) interaction models have been applied with success to model intermediate cases of micromixing. The population balance equation (PBE) approach has yielded valuable information on droplet rate processes.

It has met with limited success in analysis of both micromixing and macromixing phenomena because of the inherent difficulties of solving the multivariate PBE and incorporation of the fluid mechanics into the PBE. To this end the Monte Carlo C–D simulation techniques appear promising.

6. NOTATION

a_i	constants in Eqs. (29)
a^*	interfacial surface area per unit volume of dispersion
$A(\mathbf{x}, t)\, d\mathbf{x}$	fraction of particles at time t in the incremental range $\mathbf{x}$ to $\mathbf{x} + d\mathbf{x}$
$A_0(v, t)\, dv$	fraction of particles at time t in size range v to $v + dv$ of the feed stream
c_i	molar concentration of species i
C_{A^*}	equilibrium concentration of A at the liquid–liquid interface
$\bar{C}$	vector of molar concentrations of reactants and products
C_p	specific heat
CuR_2	copper complex
ΔH	heat of reaction
d_i	drop diameter of dispersed phase volume fraction f_i
d_{32}	Sauter mean diameter
D	molar diffusivity
$\bar{D}$	equilibrium distribution coefficient in Eqs. (25)–(28)
D_I	impeller diameter
E	enhancement factor for second-order reaction in one phase
E^*	enhancement factor for second-order reaction in both phases
E_i	enhancement factor for instantaneous reaction
f_i	volume fraction of dispersed phase, drop diameter, d_i
$f_v(y^*)\, dy^*$	fraction of the total volume of the dispersed phase with dimensionless concentration in the range y^* to $y^* + dy^*$
$g(v)$	breakage frequency of size v drop
h_A, h_B	equilibrium solubility parameter defined by Eq. (21)
$h(v, v')$	collision frequency between drops of sizes v and v' for a binary collision process based on number concentration

$h^+(\mathbf{x}; t)\,d\mathbf{x}$ number of droplets produced per unit volume of dispersion per unit time at time t in the incremental range $\mathbf{x}$ to $\mathbf{x} + d\mathbf{x}$ in phase space

$h^-(\mathbf{x}; t)\,d\mathbf{x}$ number of droplets destroyed per unit volume of dispersion per unit time at time t in the incremental range $\mathbf{x}$ to $\mathbf{x} + d\mathbf{x}$ in phase space

HR organic chelation molecule

HR_{63} aliphate oxime

HR_{65} β-hydroxy benzophenone

I ionic strength

I^* ω_i/ω_r = dispersed phase mixing modulus

k_i reaction velocity constants

$k_{LA,c}$ mass transfer coefficient of species A in phase c

$k_{LB,d}$ mass transfer coefficient of species B in phase d

K_c thermodynamic equilibrium constant based on molar concentrations

$K_{d,i}$ overall mass transfer coefficient based on dispersed phase for species i

$K_{OD,i}$ overall mass transfer coefficient, dispersed phase basis

K_r $k/\omega_r c_0$ = reaction modulus

L length of flow reactor

m equilibrium solubility parameter in Eqs. (31) and (32)

$n(c)\,dc$ fraction of drops of concentration range c to $c + dc$

$n_0(c)\,dc$ fraction of drops of concentration range c to $c + dc$ for the feed stream

$n_0(t)$ number feed rate of drops into reactor

N total number of drops per unit volume of dispersion

N_A molar flux

N_{We} impeller Weber number

Pe_α Peclet number for phase α, udL/S_α

R_i rate of production of species i due to chemical reaction

$\bar{R}_A$ apparent local rate of production of species A

s order of reaction

S_α axial dispersion coefficient, phase α

t time

t_c contact time in penetration model

T temperature

u_c, u_d superficial velocity of continuous and dispersed phases

$u_{d,i}$ velocity through the column of drop, diameter d_i

$\mathbf{v}$ velocity vector

w weight fraction of solute in continuous phase

x distance parameter in the microscopic problem

$\mathbf{x}$ general coordinate vector in particle phase space
$\mathbf{x}_e$ external coordinate vector in particle phase space
$\mathbf{x}_i$ internal coordinate vector in particle phase space
$\dot{\mathbf{x}}$ coordinate velocity in phase space
y_i weight fraction of solute in dispersed phase in ith volume fraction, f_i
y^* dimensionless concentration
z axial distance parameter in the reactor; stoichiometric coefficient for species B in equation (10)

Superscripts

α refers to continuous ($\alpha = c$) or dispersed phase ($\alpha = d$)
——— refers to organic phase

Subscripts

A refers to species A
B refers to species B
i refers to species i
0 refers to initial conditions

Greek Symbols

∇ del operator
$\beta(v', v)\,dv$ fraction of drops of volume between v to $v + dv$ produced by breakage of a drop of volume v'
δ film thickness
δ^* thickness of adsorption layer
ϵ thermal diffusivity
Γ surface concentration
$\lambda(v, v')$ collision efficiency of colliding drops of size v and v'
ν stoichiometric coefficient (moles of B to moles of A)
$\nu(v)$ number of daughter droplets produced by breakage of a drop of diameter v
ω_i dispersed phase mixing frequency
ω_r residence frequency of drops
Φ dispersed phase holdup, volume of dispersed phase to volume of total dispersion
ρ density
τ^α average residence time of phase α

REFERENCES

1. Astarita, G. (1972). Gas liquid reactors, *Chemical Reaction Engineering, Proc. 5th European/2nd Inter. Symp. on Chemical Reaction Engineering*, Amsterdam, Holland.
2. Dankwerts, P. V. (1979). *Gas Liquid Reactions*, McGraw-Hill, New York.
3. Resnick, W., and Gal-Or, B. (1968). *Advan. Chem. Eng.*, **7**, 295.
4. Rietema, I. (1964). *Advan. Chem. Eng.*, **5**, 237.
5. Tavlarides, L. L., and Stamatoudis, M. (1981). *Advan. Chem. Eng.*, **11**, 199–273.
6. Flett, D. S., and Spink, D. R. (1976). *Hydrometallurgy*, **1**, 207.
7. Ashbrook, A. W. (1975). *Coordination Chemistry Rev.*, **16**, 285.
8. Spink, D. R., and Okuhara, D. N. (1974). Comparative equilibrium and kinetics of an alkylated hydroxy quinoline and a β-hydroxy oxime for the extraction of copper, *Met. Trans. S.*, 1935.
9. Flett, D. S., Hartlage, J. A., Spink, D. R., and Okuhara, D. N. (1975). The extraction of copper by an alkylated 8-hydroxy-quinoline, *J. Inorg. Nucl. Chem.*, **37**, 1967.
10. Bauer, G. (1974). Solvent extraction of copper: kinetic and equilibrium studies, Ph.D. Thesis, Univ. of Wisconsin, Madison, Wisconsin.
11. Whewell, R. J., Hughes, M. A., and Hanson, C. (1975). Kinetics of the solvent extraction of copper II with LIX reagents: single drop experiments, *J. Inorg. Nucl. Chem.*, **37**, 2303.
12. Atwood, R. L., Thatcher, D. N., and Miller, J. D. (1975). Kinetics of copper extraction from nitrate solutions by LIX 64N, *Met. Trans. B*, **6B**, 465.
13. Miller, J. D., and Atwood, R. L. (1975). Discussion of the kinetics of copper solvent extraction with hydroxy oximes, *J. Inorg. Nucl. Chem.*, **37**, 2539.
14. Flett, D. S., Melling, J., and Spink, D. R. (1977). Reply to the discussion on the kinetics of copper solvent extraction with hydroxy oximes, *J. Inorg. Nucl. Chem.*, **39**, 701.
15. Whewell, R. J., Hughes, M. A., and Hanson, C. (1976). Kinetics of the solvent extraction of copper II with LIX reagents: the effect of LIX 64, *J. Inorg. Nucl. Chem.*, **38**, 2071.
16. Flett, D. S., Okuhara, D. N., and Spink, D. R. (1973). Solvent extraction of copper by hydroxy oximes, *J. Inorg. Nucl. Chem.*, **35**, 2471.
17. Spink, D. R., and Okuhara, D. N. (1973). *Int. Symp. Hydrometallurgy*, (Eds. D. J. I. Evans and R. S. Shoemaker), pp. 497–534, AIME, New York.
18. Flett, D. S., Okuhara, D. N., and Spink, D. S. (1973). *J. Inorg. Nucl. Chem.*, **35**, 2471.
19. Bauer, G. L., and Chapman, T. W. (1976). *Metallurgical Trans. B*, **7B**, 519.
20. Hoh, Y. C., and Bautista, R. G. (1978). *Metallurgical Trans. B*, **9B**, 69.
21. Whewell, R. J., and Hughes, M. A. (1979). The modelling of equilibrium data for the liquid–liquid extraction of metals. Part III, an improved chemical model for the copper/LIX 64N system, *Hydrometallurgy*, **4**, 109.
22. Lee, C. K., and Tavlarides, L. L. (1983). Chemical equilibrium studies on the copper-sulfuric acid-kelex 100 xylene system, *Metallurgical Trans. B.*, **14B**, 153.
23. Agarwal, R., and Tavlarides, L. L. Chemical equilibrium studies of Fe(III) sulfate–sulfuric acid/β-alkenyl-8 hydroxy quiniline–xylene extraction system, *Metallurgical Trans. B*, submitted.
24. Freeman, R. W., and Tavlarides, L. L. (1982). Study of interfacial kinetics for liquid–liquid systems—II. Intrinsic kinetics of the copper chelation reaction with an oxime-based reagent, *Chem. Eng. Sci.*, **37**, 1547.
25. Lee, C. K., and Tavlarides, L. L. Intrinsic kinetics of iron(III) sulfate–sulfuric acid/β-alkenyl 8-hydroxy quinoline-xylene extraction system, *Ind. Eng. Chem. Fund.*, to appear.
26. Cox, P. R., and Strachan, A. N. (1971). Two phase nitration of chlorobenzene, *Chem. Eng. Sci.*, **26**, 1013.

27. Hanson, C., Marsland, J. G., and Naz, M. A. (1974). Macrokinetics of chlorobenzene nitration, *Chem. Eng. Sci.*, **29**, 297.
28. Hanson, C., and Marsland, J. G. (1971). Macrokinetics of toluene nitration, *Chem. Eng. Sci.*, **26**, 1513.
29. Cox, P.R., and Strachan, A. N. (1972). Two phase nitration of toluene—I. *Chem. Eng. Sci.*, **27**, 457.
30. Cox, P. R., and Strachan, A. N. (1972). Two phase nitration of toluene—II, *Chem. Eng. J.*, **4**, 253.
31. Chapman, J. W., Cox, P. R., and Strachan, A. N. (1974). Two phase nitration of toluene, *Chem. Eng. Sci.*, **29**, 1247.
32. Harriott, A. W., and Picker, D. (1975). Phase transfer catalysis: an evaluation of catalysts, *J. Am. Chem. Soc.*, **97**, 2345.
33. Jones, R. A. (1975). Application of phase transfer catalysis in organic synthesis, *Aldrichimica Acta*, **93**, 35.
34. Dehmlow, E. V. (1975). Phase transfer catalysis, *Chem. Tech.*, **5**(4), 210.
35. Van Krevelen, D. W., and Hoftijzer, P. J. (1954). *Trans. Inst. Chem. Engs.*, **32**, S60.
36. Rod, V. (1974). Mass transfer in a heterogeneous chemical reactor with two liquid phases, *Chem. Engng. J.*, **7**, 137.
37. Mhaskar, R. D., and Sharma, M. M. (1975). Extraction with reaction in both phases, *Chem. Engng. Sci.*, **30**, 811.
38. Merchuk, J. C., and Farina, I. II. (1976). Simultaneous diffusion and chemical reaction in two-phase system, *Chem. Engng. Sci.*, **31**, 645.
39. Sada, E., Kumazawa, H., and Butt, M. A. (1977). Mass transfer with chemical reaction in both phases, *Can. J. Chem. Eng.*, **55**, 475.
40. England, D. C., and Berg, J. C. (1971). Transfer of surface-active agents across a liquid–liquid interface, *AIChE J.*, **17**, 312.
41. Yagodin, G. A., Tarasov, V. V., and Kizim, N. F. (1974). The state of substances in the organic phase and stripping microkinetics, *Proc. of Int'l. Solvent Extraction Conf.*, Sept. 8–14, Lyon, France.
42. Chapman, T. W., Caban, R., and Tunison, M. E. (1975). Rates of liquid ion exchange in metal extraction processes, *AIChE Sym. Series 152*, **71**, 128.
43. Tunison, M. E. and Chapman, T. W. (1976). The effect of a diffusion potential on the rate of liquid–liquid ion exchange, *Ind. Eng. Chem. Fund.*, **15**, 196.
44. Gal-Or, B., and Hoelscher, H. E. (1966). *AIChE J.*, **12**, 499.
45. Tavlarides, L. L., and Gal-Or, B. (1969). A general analysis of multicomponent mass transfer with simultaneous reversible chemical reactions in multiphase systems, *Chem. Eng. Sci.*, **24**, 553.
46. Tavlarides, L. L., and Gal-Or, B. (1969). Theory of coupled heat and mass transfer with zero-order reactions in two phase systems, *Israel J. Tech.*, **7**, 1.
47. Johnson, A. I., and Akehata, T. (1969). *Can. J. Chem. Eng.*, **47**, 88.
48. Ruckenstein, E., Dang, V., and Gill, W. N. (1971). *Chem. Eng. Sci.*, **26**, 647.
49. Brunson, R. J., and Wellek, R. M. (1971). *AIChE J.*, **17**, 1123.
50. Freeman, R. W. (1977). M.S. Thesis, Illinois Institute of Technology, Chicago.
51. Freeman, R. W., and Tavlarides, L. L. (1980). Study of interfacial kinetics for liquid–liquid systems—I. The liquid jet recycle reactor, *Chem. Eng. Sci.*, **35**, 327.
52. Austin, L. J., and Sawistowski, H., (1967). *I. Chem. E. Symp. Ser. No. 26.*
53. Lavis, J. B. (1954). *Chem. Eng. Sci.*, **3**, 248.
54. Brisk, M. L., and McManamey, W. J. (1969). *J. Appl. Chem.*, **19**, 109.
55. Olander, D. R., and Reddy, L. B. (1964). *Chem. Eng. Sci.*, **19**, 67.
56. Landau, J., and Chin, M. (1977). A contactor for studying mass transfer with reaction in liquid–liquid systems, *Can. J. Chem. Eng.*, **55**, 161.
57. Pavlica, R. T., and Olson, J. H. (1970). *Ind. Eng. Chem.*, **62**, 45.
58. Barnea, D. Hoffer, M. S., and Resnick, W. (1978). Dynamic behaviour of an agitated two-phase reactor with dynamic variations in drop diameter—I, *Chem. Eng. Sci.*, **33**, 205.

59. Sepulveda, J. E., and Miller, J. D. (1979). Mathematical characterization of liquid–liquid extraction in continuous flow mixers, *Paper Presented at 108th Annual AIME Meeting*, New Orleans, Feb. 18–22.
60. Gal-Or, B., and Walatka, V. V. (1967). *AIChE J.* **13**, 650.
61. Chartres, R. H., and Korchinski, W. J. (1975). Modelling of liquid–liquid extraction columns: predicting the influence of drop size distribution, *Trans. Inst. Chem. Eng.*, **53**, 247.
62. Korchinsky, W. J., and Cruz-Pinto, J. J. C. (1979). Mass transfer coefficients–calculation for rigid and oscillating drops in extraction columns, *Chem. Eng. Sci.*, **34**, 551.
63. Valentas, K. J., and Amundson, N. R. (1966). *Ind. Eng. Chem. Fund.*, **5**, 533.
64. Valentas, K. J., Bilous, D., and Amundson, N. R. (1966). *Ibid.*, 271.
65. Valentas, K. J., and Amundson, N. R. (1968). *Ibid.*, **7**, 66.
66. Verhoff, F. H. (1970). A study of the bivariate analysis of dispersed phase mixing, Ph.D. Thesis, University of Michigan, Ann Arbor.
67. Ross, S. L., and Curl, R. L. (1973). *Paper 296 4th Joint Chem. Eng. Conf.*, Vancouver, Canada, Sept. 9–12.
68. Coulaloglou, C. A., and Tavlarides, L. L. (1977). *Chem. Eng. Sci.*, **32**, 1289.
69. Ramkrishna, D. (1974). *Chem. Eng. Sci.*, **29**, 987.
70. Baipai R. K., Prokop, A., and Ramkrishna, D. (1975). *Biotechnol. Bioeng.*, **17**, 541.
71. Park, J. Y., and Blair, L. M. (1975). *Chem. Eng. Sci.*, **30**, 1057.
72. Curl, R. L. (1963). *AIChE J.*, **9**, 175.
73. Shain, S. A. (1966). *AIChE J.*, **12**, 806.
74. Hulbert, H. M., and Akiyama, T. (1969). *Ind. Eng. Chem. Fund.*, **8**, 319.
75. Evangelista, J. J., Katz, S., and Shinnar, R. (1969). *AIChE J.*, **15**, 843.
76. Baynes, C. A., and Laurence, R. L. (1969). *Ind. Eng. Chem. Fund.*, **8**, 71.
77. Ramkrishna, D. (1972). *Paper 11e Presented at 71st National Meeting*, AIChE, Dallas, Texas.
78. Shah, B. H., and Ramkrishna, D. (1973). *Chem. Eng. Sci.*, **28**, 389.
79. Hulbert, H. M., and Katz, S. (1964). *Chem. Eng. Sci.*, **19**, 555.
80. Spielman, L. A., and Levenspiel, O. (1965). *Chem. Eng. Sci.* **20**, 247.
81. Komasawa, I., Susukura, T., and Otake, T. (1972). *J. Chem. Eng. Jap.*, **5**, 349.
82. Luss, D., and Amundson, N. R. (1967). *Chem. Eng. Sci.*, **22**, 267.
83. Kattan, A., and Adler, R. J. (1967). *AIChE J.*, **13**, 580.
84. Rao, D. P., and Dunn, I. J. (1970). *Chem. Eng. Sci.*, **25**, 1275.
85. Zeitlin, M. A., and Tavlarides, L. L. (1972). *Can. J. Chem. Eng.*, **50**, 207.
86. Zeitlin, M. A., and Tavlarides, L. L. (1972). *AIChE J.*, **18**, 1268.
87. Zeitlin, M. A., and Tavlarides, L. L. (1972). *Ind. Eng. Chem. Proc. Des. Dev.*, **11**, 532.
88. Zeitlin, M. A., and Tavlarides, L. L. (1972). *Chemical reaction engineering, Proc. 5th European/2nd Intl. Symp. on Chem. Reaction Eng.*, Elsevier, Amsterdam, B1–10.
89. Ambegaonkar, A. S., and Tavlarides, L. L. (1976). *Chem. reaction engineering, Proc. 4th Int./6th European Symp. on Chem. Reaction Eng.*, DECHEMA, Frankfurt, II-76.
90. Jakubowski, S., and Sideman, S. (1976). *J. Multiphase Flow*, **3**, 171.
91. Canon, R. M., Wall, K. W., Smith, A. W., and Patterson, D. K. (1977). *Chem. Eng. Sci.*, **32**, 1349.
92. Patterson, G. K. (1981). *Chem. Eng. Commun.* **8**, 25.
93. Hsia, A. M., and Tavlarides, L. L. (1978). A simulation model for homogeneous dispersions in stirred tanks, *Paper 102F, 71st Annual AIChE Mtg.* Nov. 12–16. (1980). *The Chem. Eng. J.*, **20**, 220.
94. Bapat, P. M., Tavlarides, L. L., and Smith, G. W. (1983). Monte Carlo simulation of mass transfer in liquid–liquid dispersions, *Chem. Eng. Sci.* **38**, 2003.
95. Bapat, P. M., and Tavlarides, L. L. (1985). Mass transfer in a continuous flow stirred tank reactor. *AIChE J.*, **31**, 659.

Chapter 8

The Dispersion of Solids in Liquids

ALVIN W. NIENOW

Department of Chemical Engineering, University of Birmingham, Birmingham B15 2TT, UK

1. INTRODUCTION

This chapter refers to the dispersion of solids of such a size and concentration that they have a negligible effect on the rheological properties of the fluid. For very fine particles where interfacial phenomena may dominate the dispersion process and the resulting rheology, the book by Parfitt[1] gives an excellent introduction to the subject. In general, the solids are present as a collection of individual particles and not as flocs or weak agglomerates whose stability is affected by the agitation. Of course, some abrasion due to particle–particle[2] and impeller–particle[3] impacts may occur but the size of the parent particles is not materially affected.

With the restrictions listed above, the processes taking place could include crystallisation and precipitation, dissolution and leaching, ion-exchange and adsorption and catalysed chemical reactions. In certain cases, gases may be evolved by reaction (evaporative crystallisation) or they may be introduced as reactants (catalysed hydrogenations). Thus solid suspension in the presence of gas is considered. In addition, solids must often be withdrawn continuously usually with a required residence time distribution in crystallisation and leaching operation for example.

In all these processes, the impeller Reynolds number, ND^2/ν, is generally greater than 2×10^4 so that the system can be considered fully turbulent. Though a wide range of vessel geometries can be conceived, a standard geometry has generally been investigated in

273

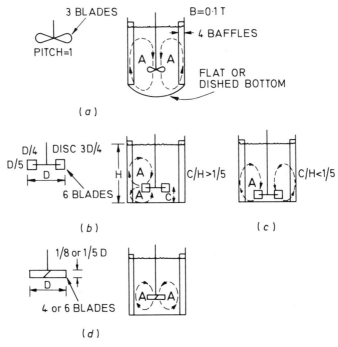

Figure 1   Turbulent flow patterns in baffled vessels. (a) with a propeller; (b) and (c) with disc turbines; (d) with a 45° pitched-blade turbine pumping downward.

the literature. These vessels have had an equal liquid height-to-diameter ratio with four, vertical strip baffles 10% of the vessel diameter to aid vertical flow of the liquid and prevent aeration due to vortexing. The base is usually flat or dished (see Fig. 1). Variations from this basic configuration will be pointed out as appropriate.

2.   PARTICLE SUSPENSION AND FLUID MECHANICS

2.1.   Interaction between the Particles and the Fluid

The flow in mechanically-agitated vessels is quite markedly three-dimensional and, under the conditions defined in the Introduction, highly turbulent. It is also far from homogeneous with the highest energy dissipation rate occurring in the region of the impeller[4]. In parts of the vessel, near the centres of the circulation loops (marked A in Fig. 1), the intensity of turbulence may be greater than 100%[5].

The difficulty of analysing particle dispersion in such a complicated flow field is compounded by the fact that two separate regions must be considered. Firstly, particles must be lifted from the base and secondly, once lifted, they must be dispersed into the remainder of the vessel. In general, the mechanisms that gives rise to lift are dependent on the fluid flow only at the base of the vessel whilst the mechanisms giving rise to dispersion are dependent on the flow throughout. Insufficient information is known at present to separate these two regions.

Visual observations of the base of a transparent agitated vessel suggest two mechanisms having an effect there. Firstly, the flow at the base is either swirling outwards (Fig. 1a, 1c, or 1d) or inwards (Fig. 1b). If the flow is of sufficiently high velocity, particles are moved by it; in the former cases, towards the periphery of the vessel and in the latter, towards the centre. This movement is related to a balance between the drag and lift forces arising from the hydrodynamic regime around the particle and the gravitational forces associated with it. The fluid flow can be thought of as a boundary layer one across the base of the tank which, depending on the size of the vessel, may remain entirely laminar[6] or become turbulent somewhere along its length. Either way, the velocity in this region will increase rapidly in a vertical direction from zero at the base and in consequence, the velocity at the centre of mass of particles of increasing size also grows rapidly.

Thus, it might be expected that the boundary layer mechanism causing particles to be swept across the base would be relatively insensitive to particle size (unlike terminal velocity in a still fluid) and particle suspension is found to be so. However, lift is very dependent on particle shape and extreme shapes may either increase or decrease the ease of particle suspension[7,8].

During the process of sweeping across the base and particularly once small piles (or fillets) have been formed, particle pick-up can be clearly seen to be caused by turbulent bursts[7,9] such as indicated by Fig. 2[10]. These bursts originate in the turbulent mainstream bulk flow above the base. However, whilst turbulence burst theory has been applied to sub-micron particles with some success, it has not proved possible to do so for particles of the sizes normally of interest in stirred vessels[9]. However, an indication of the importance of these turbulent bursts in lifting particles from the base can be obtained from the study of Ohiaeri[11]. Using a downward facing laminar jet to simulate flow from a propeller (Fig. 1c), he found much higher power levels were necessary to suspend particles with the jet than with a rotating propeller.

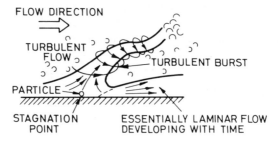

Figure 2 Schematic diagram of a turbulent burst at the base[10].

Whilst little is known of the detailed flow close to the base and the way it interacts with particles thereon, much more is known of the main flow (see Chapters 2 and 3). As in all turbulent flows[12], the fluid contains large, energy containing eddies of size of the order of the system size. These give rise to smaller eddies by energy transfer and all of these large eddies are anisotropic. However, these give rise to smaller eddies which are isotropic. There is then a spectrum of eddy sizes down to the Kolmogoroff scale through which energy passes but without dissipation. This is known as the inertial sub-range. At the Kolmogoroff scale (eddy size, λ_K), inertial and viscous forces are in balance and by definition, the eddy Reynolds number, $Re_K (= \lambda_K u_K'/\nu) = 1$ where u_K' is the velocity of an eddy of size, λ_K. Below the Kolmogoroff scale ($\sim 50\,\mu$m), energy is dissipated by viscosity in the viscous sub-range.

Small scale isotropic eddies in the inertial sub-range of sizes are only dependent on energy dissipation and kinematic viscosity. If they were controlling particle suspension, it would be expected that all impeller types would be equally effective, that is, need the same energy dissipation rate at particle suspension. In fact, this is not the case and very considerable differences in the energy dissipation rate are required by different impellers. These differences indicate that it is the mean convective flows and large anisotropic eddies which are most important.

The importance of the larger scale eddies can be further appreciated from the work of Quraishi et al.[13]. They studied particle suspension in water and water containing drag reducing agents (DRA). DRAs reduce the small scale turbulent eddies but have little effect on the large scale convective flows. They found that the stirrer speed required for particle suspension was unaffected by the presence of DRA. Thus, it can be concluded from their work too that it is the large scale flows which are of most importance. This implies that some optimum geometry should exist for particle suspension, that is, one which would minimise energy dissipation rates for a particular suspension requirement. The work of Aesbach and Bourne[14] supports this too.

Another factor requiring consideration is the effect of particle concentration. If there are more particles present at the base, more turbulent bursts are required to suspend them, that is, the energy of a particular burst will entrain a certain number but will leave others behind until another burst arrives. Once in suspension, the presence of particulate solids has a damping effect on turbulence and thereby reduces the formation of the turbulent bursts which are required to sustain suspension. Thus, qualitatively, it can be seen that increasing solids concentration requires increasing energy for suspension.

It has been proposed that, above about 17% by volume of solids, no further increase in agitator speed is required because a hindered settling regime is entered[15]. However, the little work that has been conducted at these concentrations is inconclusive. The conditions are then approaching those where the rheological characteristics of the fluid–solid mixture must be taken into account. These conditions are outside the scope of this chapter but Chapter 4 gives some insight.

One final point is worthy of note here. The power drawn by the impeller is critically dependent on the flow close to the impeller blades, especially the trailing vortex systems associated with them[16-18]. This trailing vortex system rotates at a high velocity and in consequence is a low pressure region. It therefore attracts gas to it which generally leads to a considerable drop in power even for low gas hold-ups[19]. However, in solid–liquid systems, the high speed rotation due to the impeller itself and the vortices tends to centrifuge particles out of this region. As a result, the power drawn by a rotating impeller increases much more slowly than the increase in bulk density. Indeed, Chapman et al.[8] found no increase in power number up to 20% by weight solids when using a disc turbine agitator (Fig. 1b).

2.2. Models of Particle Suspension

Because of the complexity of the fluid mechanics only simple, rather qualitative models have been developed.

The model of Nienow and Miles[20] assumes that particle suspension is dependent on the convective flows near the base. In this region, a minimum average velocity independent of direction, $\bar{u}_{base}$, is necessary to achieve complete particle suspension. However, the velocity at any level, $\bar{u}$, in an agitated vessel based on many experimental studies[5, 21, 22] is given by

$$\bar{u} = A \cdot (ND^2)/(T^2H)^{1/3} \tag{1}$$

where A is dimensionless coefficient of proportionality which decreases rapidly with increasing distance from the impeller. The first conclusion from this model then is that the impeller should be placed close to the vessel base in order to achieve $\bar{u}_{base}$ with the lowest agitator speed for a particular type and size of agitator and vessel. Experimental results support this.

Secondly, the power drawn by the impeller is given by

$$P = Po\rho_L N^3 D^5 \tag{2}$$

where Po is the power number and the mean energy dissipation rate, ϵ_T, is given by

$$\epsilon_T = P/\rho_L V \tag{3}$$

For fully baffled, fully turbulent flows, Po is constant[23], dependent only on the impeller type and vessel geometry. Thus since the vessel volume is proportional to T^2H, then at any level, including the base,

$$\bar{u} \propto \bar{u}_{base} \propto \epsilon_T^{1/3} D^{1/3} \tag{4}$$

Therefore, it follows that $\bar{u}_{base}$ is obtained at a lower value of energy dissipation rate by using a large size of impeller. Alternatively, large slow impellers should enable particle suspension to be achieved at lower values of ϵ_T in the same size of vessel as compared to small ones of the same type. This is in agreement with experiment for radial flow agitators[7, 24] but not for propellers.

Equation (4) also implies that a lower energy dissipation rate is required in large scale systems compared to geometrically similar small scale ones. This qualitative conclusion is in agreement with all experimental studies where energy dissipation rate has been measured on different scales and with the design guide of one mixer manufacturer[25].

The other successful model is that of Baldi et al.[26] This approach indicates much more the significance of the particle and fluid physical properties. However, since it is linked to mean energy dissipation rates and Kolmogoroff's theory of isotropic turbulence, it is somewhat less realistic from the fluid mechanics point of view. They assumed that particles were picked up by turbulent eddies of a critical size. This size, λ, would be of the order of the particle size because smaller ones would be insufficiently energetic to achieve suspension; and the low frequency of arrival of larger eddies would reduce their chances of achieving it. An energy balance can then be made between the kinematic energy assuming that a particle must be lifted to a height, d_p, in order to become entrained. Thus

$$\rho_L u'^2_{\text{base}} \propto d_p \Delta \rho g \tag{5}$$

In general, these eddies are much larger than the Kolmogoroff scale of turbulence, λ_K. In the inertial sub-range, the fluctuating velocity, u' can be expressed as

$$u'_{\text{base}} \propto (\lambda (\epsilon_T)_{\text{base}})^{1/3} \propto (d_p (\epsilon_T)_{\text{base}})^{1/3} \tag{6}$$

where $(\epsilon_T)_{\text{base}}$ is the local energy dissipation rate at the vessel base. Assuming $(\epsilon_T)_{\text{base}} \propto \epsilon_T$ and since

$$\epsilon_T = (4 \, \text{Po} \, N^3 D^5)/\pi T^3 \tag{7}$$

for cylindrical vessels of height equal to diameter, then at the agitator speed to just cause complete suspension, that is, at $N = N_{JS}$,

$$\sqrt{\frac{g\Delta\rho}{\rho_L}} \cdot \frac{T d_p^{1/6}}{\text{Po}^{1/3} D^{5/3} N_{JS}} = \text{const.} = Z \tag{8}$$

Alternatively,

$$N_{JS} = \left(\frac{g\Delta\rho}{\rho_L}\right)^{1/2} \cdot \frac{1}{\text{Po}^{1/3}} \cdot \frac{T}{D} \cdot \frac{d_p^{1/6}}{D^{2/3}} \cdot Z \tag{9}$$

Measurements of the speed required to just completely suspend particles, N_{JS}, has shown that it is related to particle size, liquid and solid density in approximately this fashion and to be almost independent of viscosity (see below).

However, Baldi et al.[26] also considered that Z is a parameter to be found by experiment and that the relationship between $(\epsilon_T)_{\text{base}}$ and ϵ_T actually depends on the fluid properties especially viscosity, and on the impeller size, position and type. Therefore

$$Z = f(\text{Re}^*; T/D; C/D; \text{impeller type}) \tag{10}$$

where $\text{Re}^* = \rho_L D^3 N_{JS}/\mu_L T$; Re^* can be considered as a Reynolds number for suspension. Recent experimental work[8] has shown that though this approach works well for disc turbine impellers, confirming the experimental results of Baldi et al.[26], it is less successful for other geometries.

The probable reason for the inadequacy of the models is because the flow is too complex to be analysed by any simple one. Pipeline flow, due to its well-defined geometry, is inherently simpler but even there, in order to obtain correlations which are applicable over a wide range of flow conditions, four regimes have been identified[27]. These workers also develop a model for the critical velocity for solid transport in a pipeline which is equivalent to N_{JS} here. They need to invoke a parameter which is the fraction of eddies with a velocity greater than the still fluid terminal velocity of the particle to achieve a satisfactory model. Such a parameter is not available for the base region of a stirred vessel. Finally, after developing a mathematical model, Oroskar and Turian[27] abandon it for an empirical equation based on a regression analysis of experimental data.

If further advances are to come in modelling particle suspension in stirred vessels, it will do so as a result of computer modelling of the whole tank flow field; and then applying that flow structure to the solid suspension problem. Work along these lines is in progress[28, 29].

3. STATES OF SUSPENSION

3.1. Just Completely Suspended

This is the state most commonly studied and of greatest importance. It corresponds to the lowest agitator speed, N_{JS}, or specific power input, $(\epsilon_T)_{JS}$, at which all of the surface area of the particles is available for processing. Many workers[30, 31] have shown that the rate of solid–liquid mass transfer increases relatively rapidly with increase in agitation speed or power input up to this point but only slowly above it.

To try to quantify a little the importance of this relative insensitivity to agitation conditions above N_{JS}, consider the two extreme cases of solid–liquid reactions[32, 33]. For reactions where the rate limiting step is bulk diffusion, that is, mass transfer limited, then typically a ten-fold increase in power above $(\epsilon_T)_{JS}$ only increases the rate by about one third. Of course, when the rate limiting step is at the surface, for example, surface integration in crystallisation or in-pore diffusion for catalyst powders, increases in power do not cause any increase in overall reaction rate. The effect of agitation conditions on mass transfer rates is discussed in more detail in the last section of this chapter.

N_{JS} is usually determined visually as the speed at which particles do not remain stationary on the base for more than 1 to 2 seconds[24]. The mean energy dissipation rate or power input may then be estimated if the impeller power number, Po, is known. However, suitable instrumentation enables $(\epsilon_T)_{JS}$ to be measured directly. This is to be preferred, especially on the small scale, since even minor changes in impeller geometry can have a profound effect on the power number[34].

Instrumented and sampling techniques have also been developed which are less subjective. They are based on determining the impeller speed, N_{PJS}, for which the particle concentration just above the base of the vessel is a maximum[35] or shows a discontinuity[36] with the impeller speed. The estimation of this speed involves measuring local solids concentration in situ or by withdrawing a sample from the vessel. Bourne and Sharma[35] showed experimentally that for a wide range of geometries and particulate solids, a peak occurs when a graph of solids concentration in the withdrawn sample is plotted against impeller speed (Fig. 3), provided the sample was withdrawn from near the base of the vessel.

The maximum shown in Fig. 3 can be explained as follows. At low speeds, the majority of the particles rests on the bottom, producing a very low concentration reading just above the base. As speed is

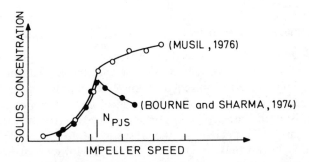

Figure 3   Particle concentration near the base vs. impeller speed: Determination of N_{PJS} by a peak or a kink (diagrammatic).

increased so the particles are gradually suspended and hence the local concentration just above the base increases. Eventually a speed is reached when the source of particles on the base is practically exhausted but the upper regions of the dispersion have very low particle holdups. Thus further speed increases tend to homogenise the suspension and therefore reduce the local concentration near the base, producing a maximum as shown in Fig. 3. However this rather idealised description does not always apply. Musil[36] also noted that occasionally a "kink" or discontinuity, in the form of a sharp decrease in gradient rather than a peak, would occur.

Bourne and Sharma[35] found that N_{JS} and N_{PJS} coincided. As well as being less subjective, the technique for determining N_{PJS} also offers advantages in large scale and three phase (gas–solid–liquid) studies where visual observation is difficult or impossible. Initial tests on three phase systems[9] also indicated agreement between the visual and instrumented techniques. However, more extensive measurements in vessels from 0.29 to 1.8 m diameter[8] have shown discrepancies. The same instrumentation methods can be used in three phase and large scale systems to give a suspension speed. Indeed, the values from the two techniques often coincide or are very close. However, in general, the instrumented technique gives lower agitation speeds than the visual one in both two phase and three phase systems, that is, $N_{PJS} \leqslant N_{JS}$. In retrospect, the reasoning is rather simplistic which suggests N_{PJS} and N_{JS} should precisely coincide.

Because of the majority of work in the literature has measured N_{JS} and its physical meaning and its link to reaction rates is understood, the subsequent parts of this chapter concentrate on correlations for N_{JS}.

3.2. Homogeneous Suspension

This condition is defined as that which exists when the particle concentration and size distribution is constant throughout the vessel. The minimum speed required to achieve it, N_{JH}, is generally very much greater than N_{JS} and so too therefore is the power. This criterion is often stated to be important for vessels from which solids are discharging, especially if the vessels are to be modelled as continuous stirred tank reactors (C.S.T.R.'s). However, it has been clearly shown that the method of solid withdrawal is as important as the state of suspension[14]. If severe classification by size or density is to be prevented, isokinetic withdrawal or a good approximation to it is required.

Aesbach and Bourne developed the concept of wash-out curves to indicate the achievement or otherwise of C.S.T.R. behaviour. Since then various workers[37, 38] have developed the concept of using a steady-state separation coefficient K where

$$K = C_{out}/\bar{C} \tag{11}$$

C_{out} is the solids concentration in the outflow from the vessel and $\bar{C}$ is the mean concentration within it. K itself depends on the geometry, the solid and liquid properties and the withdrawal method as well as the agitation level, and K is equal to 1 if the vessel exhibits C.S.T.R. behaviour. If a quantity of solids is added to the vessel in an unsteady state wash-out test to give an initial concentration in the vessel $\bar{C}(0)$, then it can be shown[38]

$$C_{out}(t)/\bar{C}(0) = K \exp \{-t/(\tau_L/K)\} \tag{12}$$

or

$$C_{out}(t)/C_{out}(0) = \exp (-t/\tau_S) \tag{13}$$

where τ_L and τ_S are the mean residence time of the solid and liquid respectively. Thus K and τ_S can be obtained. The use of K has been demonstrated for solid–liquid chemical reactors[39] and for crystallisers[6].

The inclusion of a draught tube in conjunction with an axial flow impeller greatly enhances the likelihood of achieving high K values and on a small scale, a value of 1 is possible[14]. The same workers have shown that deviations from co-linearity lead more easily to reductions in K (and therefore increases in τ_S) than inadequate matching of velocities. With radial flow impellers, K values of 1 have been achieved on a very small scale by intermittent withdrawal[40].

An example of where homogeneous suspension within a vessel may be desirable in large vessels is in crystallisers where the presence of crystals in regions of high supersaturation may prevent primary heterogeneous nucleation. However, if this is a problem[41], the agitation levels required to prevent undesirable liquid inhomogeneities will be the governing ones.

3.3. Interface Height Equal to 90% of the Liquid Height

In many cases, the liquid in the upper regions of a vessel is without solids at N_{JS}[7]. For this reason Mersmann et al.[42] and Einekel and Mersmann[15] suggested that a minimum suitable agitation speed should be one which ensures particles reach to within 10% of the upper interface, N_{90}. However, this definition also has disadvantages. Firstly, for fine and light particles, once they are lifted from the bottom, they easily reach the top of the liquid. In this case therefore, N_{90} is less than N_{JS} and by working at N_{90}, some solid surface is not available for processing. Secondly, for large and/or dense particles, N_{90} is greater than N_{JS}. However, the liquid is generally well mixed even at N_{JS} and, provided all the particles are suspended, each sees on average the same liquid environment. Thus, extra power is expended for no return in more efficient liquid–solid contacting.

In passing, it should be noted that the use of multiple impellers in tanks of H/T ratio greater than 1 helps disperse solids more readily throughout the fluid but only at the expense of very enhanced power levels, i.e., N_{JS} for two impellers is approximately equal to N_{JS} for one whilst P_{JS} for two impellers is approximately equal to twice P_{JS} for one[8]. On the other hand, gas sparging greatly increases the solids level and, if done specifically for this purpose into the upper part of the vessel, with little or no extra power consumption[43]. The effect of agitation on solids level has also been studied by Weismann and Efferding[44].

4.   THE EFFECT OF PHYSICAL PROPERTIES AND GEOMETRY ON N_{JS}

4.1.   Standard Geometries

In spite of much work over the last quarter century, the best overall correlation is still that put forward first by Zwietering[24]. He conducted over a thousand experiments with a wide range of impeller types in vessels up to 0.7 m diameter and interpreted the results using dimensional analysis. Expanding the dimensionless groups gives the following purely empirical expression:

$$N_{JS} = S\nu^{0.1}d_p^{0.2}(g\Delta\rho/\rho_L)^{0.45}X^{0.13}D^{-0.85} \qquad (14)$$

The similarity with Eq. (9) can be clearly seen.

The exponents on ν, d_p, $(g\Delta\rho/\rho_L)$, D and X were found to be independent of impeller type, vessel size, impeller clearance and impeller to tank diameter ratio. A review of many other workers correlations indicates that very similar exponents have generally been found[32] and two recent studies[8, 11] in tanks from 0.29 to 1.8 m diameter are also in reasonable accord with them.

S is a dimensionless particle suspension parameter and data are available in graphical form (Fig. 4) for a variety of impeller types, impeller clearances and impeller to tank diameter ratios[7, 24] Zwietering used two flat paddles, a vaned disc and a rather primitive

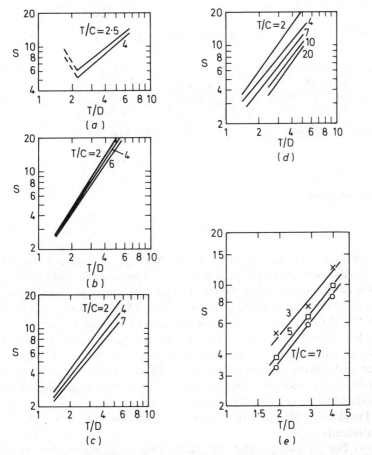

Figure 4   S-values for different impellers: (a) Propeller; (b) Vaned disc; (c) Flat paddle (2 blades, width 0.5 D); (D) Flat paddle (2 blades, width 0.25 D); (e) Disc turbine. (a to d after Zwietering[24]; e after Nienow[7]).

propeller and Nienow a disc turbine. More S values are given by Nienow and Miles[20] and Chapman et al.[8]

In general,

$$S \propto (T/D)^a \qquad (15)$$

where $a = 0.82$ for propellers[24] and about 1.3 for radial flow impellers[7, 24]. Since

$$N_{JS} \propto SD^{-b} \qquad (16)$$

where $b = 0.85$, then

$$P_{JS} \propto \text{Po } N_{JS}^3 D^5 \propto \text{Po } S^3 D^{5-36} \qquad (17)$$

or for one size of vessel

$$P_{JS} \propto \text{Po } D^{5-3(a+b)} \qquad (18)$$

Since, in addition, for geometrically similar systems,

$$(\epsilon_T)_{JS} \propto \text{Po } N_{JS}^3 D^5 / V \propto \text{Po } N_{JS}^3 D^2 \qquad (19)$$

then on scale up,

$$(\epsilon_T)_{JS} \propto \text{Po } S^3 D^{2-3b} \qquad (20)$$

Of course, because power is proportional to N^3, small differences in measured values of S, of a and of b can have a profound effect on each of the above relationships. In addition, errors in the measurement of Po and, if taken from the literature, variations of Po with D/T and C/T ratios as well as with scale which may not be allowed for has caused a very wide range of differing results to be published. The conclusions below are therefore, based on the most extensive and authorative experimental studies. In addition, they either agree qualitatively with the fluid mechanics (as it is understood) or the models that have been discussed.

Provided strict geometric similarity is maintained, the following statements apply:

(a) For propellers and 45° pitched-blade agitators, the power required for particle suspension is relatively insensitive to relative impeller size[8, 24]. (Also from Eq. (18), with $a = 0.82$, $b = 0.85$.)

(b) For radial flow impellers, large slow moving ones require less power than small fast ones[20, 24]. (Also from Eq. (18), with $a = 1.3$, $b = 0.85$.)

(c) Reduction in impeller clearance, reduces N_{JS} and $(\epsilon_T)_{JS}$ except with T/D ratios less than or equal to 2 and C/H ratios less than or equal to 1/6 when dead zones beneath the impeller can occur[20, 24]. When H/T is greater than 1, it is recommended that impeller clearance should be determined as C/T rather than C/H.

(d) Increasing the number of blades on any type of agitator reduces the power required[8, 45]. The increase in Po is more than offset by the reduction in S.

(e) There is a distinct ranking order of impellers with some requiring considerably less power to suspend solids than others. Table 1 gives some typical values[8]. Similar results were obtained by Nienow and Miles[20] and Rieger et al.[45]. In general, high-flow, low-head impellers of which the propeller is the prime example, require less power than high-shear, low-flow ones of which the disc turbine is the most extreme. However, the direction of pumping is also important and pumping upwards so that considerable energy is dissipated from the discharge stream before it moves into the lower part of the vessel is clearly inefficient (Table 1).

(f) Whenever the energy dissipation rate has been measured on different scales[9, 11, 20], suspension has been achieved on the large scale at lower specific power than on the small. The relationship arising from Zwietering's work (Eq. (20), $b = 0.85$) is

$$(\epsilon_T)_{JS} \propto D^{-0.55} \qquad (21)$$

This appears to imply too little power on the large scale and the

Table 1   Comparison of Impeller Types for Particle Suspension in a 0.56 m Diameter Vessel (Chapman et al.[8])

Impeller Type	N_{JS} rev/s	$(\epsilon_T)_{JS}$ W kg^{-1}	Po (—)
3-bladed marine propellers	3.55	0.33	0.5
4-bladed, 45°-pitch pumping downwards	2.75	0.36	1.4
6-bladed disc turbine	2.02	0.61	5.9
4-bladed, 45°-pitch pumping upwards	3.6	0.72	1.2

($H = T$; $C = T/4$; $D = T/2$; $X = 3\%$; $\rho_s = 2480$ kg/m^3; $d_p = 206$ μm; distilled water at 25°C)

recent measurements of $(\epsilon_T)_{JS}$ by Chapman et al.[8] for vessels from 0.29 to 1.8 m diameter show that

$$(\epsilon_T)_{JS} \propto D^{-0.28} . \tag{22}$$

This is in good agreement with a manufacturers scale-up rule[25]; and with the recent extensive work of Einenkel and Mersmann[15] (which, however, gives different absolute values because of the different suspension criterion). For standard geometries, two approaches to scale-up are recommended. Firstly, small scale tests measuring N_{JS} and $(\epsilon_T)_{JS}$ with the impeller of interest are made and scale-up is then done using Eq. (22). Alternatively, N_{JS} (and hence $(\epsilon_T)_{JS}$) can be calculated from Eq. (14) for the small scale and the power on the large scale estimated by the same scale-up equation.

4.2. Non-Standard Geometries

The shape of the bottom of the vessel can have a profound affect on the ease of achieving particle suspension. Though Zwietering[24] found similar results for flat and shallow dished bottoms, conical ones which aid solid withdrawal make particle suspension extremely difficult[36]. The use of a hemispherical bottom causes similar problems[11]. However, even with flat bottoms, recent work[8] has shown that the presence of sparge pipes close to the surface increases both N_{JS} and $(\epsilon_T)_{JS}$, the latter by as much as 30%.

On the other hand, the use of axial flow or pitched blade impellers with a draught tube can considerably reduce both N_{JS} and $(\epsilon_T)_{JS}$ provided the flow area between the tube and the base and between the inside and the outside of the tube are all kept the same. If constrictions are caused by it, which could easily happen with hemispherical bottoms, a marked increase in $(\epsilon_T)_{JS}$ is found in spite of a reduction in N_{JS}[46]. If the base is also contoured in such a way (Fig. 5) that all dead zones are eliminated, then a minimum energy requirement for particle suspension is achieved[14].

This geometry is ideal for such operations as crystallisation on the small scale, but it may be uneconomic to fabricate on the large scale. However, with bottom–entry impellers, the stuffing box could be suitably shaped. Indeed, in recent work in vessels of standard geometry of 0.30 m and 0.56 m diameter, Chapman et al.[8] have shown that just the presence of a bottom–steady bearing which effectively removes the dead zone beneath the impeller can markedly reduce N_{JS} and $(\epsilon_T)_{JS}$. For example, in a 0.3 m vessel, the

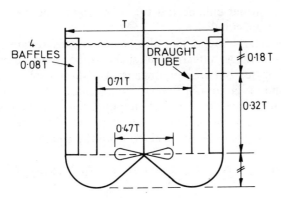

Figure 5   The geometry for minimising $(\epsilon_T)_{JS}$: contoured bottom and equal flow areas[14].

change from the standard geometry of Fig. 1b to an identical one with a bottom–steady reduced $(\epsilon_T)_{JS}$ by 35%.

The dramatic affect of a sparge pipe and a bottom–steady bearing on $(\epsilon_T)_{JS}$ seems to offer an additional explanation why such a wide variation in specific power requirements on scale-up have been found. The extreme sensitivity to geometry reinforces the difficulties set out in Eqs. (15)–(20) and the subsequent discussion.

4.3.   Size Distribution and Shape

Little has been done on these aspects. Baldi et al.[26] and Chapman et al.[8] using two, three, and four component mixtures found that the use of a mass mean enabled N_{JS} for the mixtures to be correlated along with data from close-sieved particles.

Nienow[7] used a wide range of shapes and provided the solids were granular found shape to be relatively unimportant. Chapman et al.[8] found no effects for particles with sphericity from 1 to 0.7. However, 550 μm anthracite in the form of platelets (sphericity, 0.4–0.5) required almost double the impeller speed to suspend them that was expected by comparison with the other materials.

5.   THE DISPERSION OF FLOATING SOLIDS

This problem has rarely been studied. This process is generally more energy-intensive than solids suspension and is complicated by the fact that if the solids are fine, they may contain large amounts of entrapped air thereby reducing their effective density. This latter difficulty is further enhanced if the solids cannot be wetted or only wetted with difficulty[11].

Joosten et al.[47] found that the central vortex obtained without baffles was not as effective at sucking the solids down as an off-centre vortex obtained by means of a short baffle (see Fig. 6). They found a large $(D/T = 0.6)$ four-bladed, 45°-pitch impeller placed near the base of the vessel required the minimum power and that for vessels from 0.27 to 1.8 m diameter, the minimum speed required, N_{DF}, could be determined from the equation

$$(Fr)_{DF} = N_{DF}^2 D/g = 3.6 \times 10^{-2}(D/T)^{-3.65}(\Delta \rho / \rho_L)^{0.42} \qquad (23)$$

with particle concentration and size having a negligible effect. The particle size was between 2 and 10 mm and therefore air entrapment was not a problem.

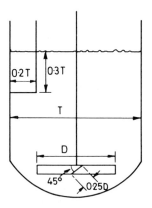

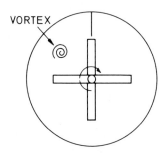

Figure 6.   The preferred baffle configuration and the position of the vortex (approximately to scale)[47].

6. PARTICLE SUSPENSION IN GASSED SYSTEMS

6.1. Introduction

Three phase slurry reactions are often carried out in stirred vessels and though the reaction kinetics are often studied, very little consideration is given to the hydrodynamics involved. Even a recent review article[48] quotes Zwietering's correlation, Eq. (14), for N_{JS} without suggesting gassing might invalidate it. On the other hand, the theoretical relationship, Eq. (9), of Baldi et al.[26] implies that N_{JS} must be increased since the gassed power number, $(Po)_g$, is less than the ungassed, Po, and Z is approximately constant. In addition, early work on three phase systems[49] suggested that at a critical gassing rate, a catastrophic total collapse of suspended particles could occur when using shrouded impellers in froth flotation cells. Clearly, the hydrodynamics need to be considered with some care and work on three phase systems has recently grown apace[9, 43, 50-52].

Until recently, impellers which require the least energy for particle suspension, that is, axial, and 45° pitch impellers as indicated in Table 1, have been considered to be unsuitable for gas dispersion. For the latter use, disc turbines are commonly employed though these require relatively high levels of power consumption for particle suspension. For three phase systems since, a priori, it is not clear which duty, that is, gas or solids dispersion, is the most difficult, a whole range of impeller types needs to be considered anew. In addition, the effect of the physical properties and particle concentration as implied by Eq. (14) needs to be known.

No detailed discussion on gas dispersions is presented here as this phenomenon is considered in Chapter 6. However, reference should also be made to the work of Chapman et al.[53] for gas dispersion mechanisms for impellers other than disc turbines.

6.2. Downward Pumping Impellers[43]

These are susceptible to severe instabilities under gassed conditions. These arise from the opposing effects of gas buoyancy upwards and impeller thrust downwards so that the flow patterns are either dominated by one or the other. If at an impeller speed just sufficient to dominate the overall flow pattern (which corresponds closely to the minimum speed required to just disperse the gas, N_{CD}), a small increase in gas rate occurs, an immediate switch from one type of flow pattern to the other arises accompanied by fluctuations in power and torque by as much as a factor of 2. This problem appears to be universal for all downward pumping impellers in gassed systems[54, 55] but is worse for small D/T ones.

For particles of size $>80 \mu$m and density ≥ 1200 kg/m³ in water, the minimum agitator speed to just suspend the solids under gassed conditions, N_{JSg}, is always greater than or equal to N_{CD} (Fig. 7). However at low gas rates (vvm* $\leq 1/4$), the energy dissipation rate required to suspend the solids, $(\epsilon_T)_{JSg}$, is considerably less than that required by disc turbines (See Fig. 8). As the gas rate is increased, very

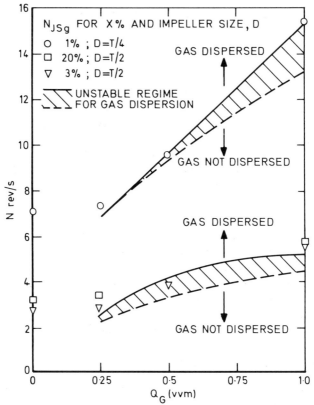

Figure 7   Comparison of gas dispersion and particle suspension conditions for a 4-bladed 45°-pitch impeller (downward pumping) in a 0.56 m diameter vessel ($\rho_s = 2500$ kg/m³; $d_p = 206 \mu$m; $H = T$; $C/H = 1/4$; distilled water at 25°C)[43]

---

*vvm is the specific gassing rate in (volumetric flow rate of gas/minute)/volume of liquid in the vessel. It is expressed here in this way for two reasons. Firstly, it compensates for the effect of scale of equipment. Secondly, values of 1/4 to about 2 vvm are commonly used and are easy to remenber.

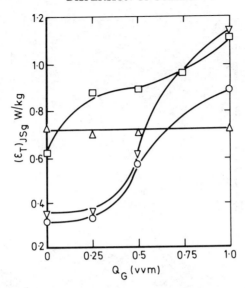

Figure 8 Specific power input required to suspend solids vs. gas rate (expressed as vvm) for various impellers in a 0.56 m diameter vessel ($H = T$; $C = H/4$; $D = T/2$; $X = 3\%$, $\rho_s = 2500\ kg/m^3$, $d_p = 206\ \mu m$; distilled water at 25°C; $\square$, 6-bladed disc turbine; $\bigcirc$, 3-bladed marine propeller; $\triangle$, 4-bladed, 45°-pitch impeller (upward pumping); $\triangledown$, 4-bladed, 45°-pitch impeller (downward pumping))[43].

substantial increases in $(\epsilon_T)_{JSg}$ and N_{JSg} are required (see Fig. 9 where $\Delta N_{JS} = N_{JSg} - N_{JS}$). In spite of the sensitivity of N_{JSg} to gas rate and of the instabilities, the catastrophic collapse of suspension reported by Arbiter et al.[49] is not observed except with the small $1/4$ D/T, 45° pitch impeller. In all other cases, a relatively gradual fall-out of particles occurs[56] and this observation is in agreement with other recent reports.

6.3.   Upward Pumping 45°-Impeller[43]

With an upward pumping impeller, the momentum of the two-phase flow resulting from the sparged gas acts in the same direction. Thus there is no instability though a minimum impeller speed, N_{CD}, is required to disperse the gas into the lower part of the vessel. Again N_{CD} is less than N_{JSg} except for particles of density only very slightly greater than that of the fluid. Because the impeller flow and the two-phase momentum act together, increases in gassing rate cause only a small increase in N_{JS} and $(\epsilon_T)_{JSg}$ is almost independent of gassing rate (Fig. 8). Thus, though under ungassed conditions, the

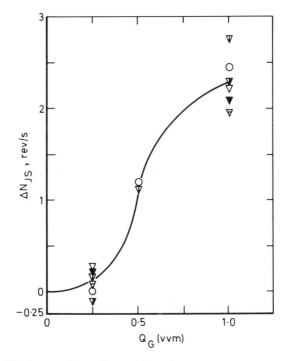

Figure 9   The increased impeller speed required to suspend particles, ΔN_{JS}, with increased gassing rate, Q_G (expressed as vvm), for propeller and 4-bladed 45°-pitch impellers pumping downwards ($D = T/2$; $C = T/4$; $1\% \leqslant X \leqslant 20\%$; $0.29\,\text{m} \leqslant T \leqslant 0.91\,\text{m}$; $\rho_s = 2500\,\text{kg/m}^3$ in water)[43].

upward pumping impeller is least efficient, its relative efficiency improves with increasing Q_G until at 1 vvm, it requires least power.

6.4.   6-Bladed Disc Turbines

This type of impeller is probably the best documented of all and was the most extensively tested by Chapman et al.[43]. As in ungassed systems, a lower clearance and large impeller give solid suspension at the lowest power. However at $C = T/6$, instabilities in flow pattern result. A single circulation loop is found at low gas rates as it is under ungassed conditions (Fig. 1c). However at higher gas rates, the figure-of-eight is found which under ungassed conditions is associated with higher clearances (Fig. 1b). On the other hand, at $C = H/4$, no instability is found.

Thus, since a clearance of $H/4$ is very suitable for gas dispersion and large disc turbines ($D = T/2$) are both the most energy efficient for this purpose[57] and, as explained earlier in this chapter, for particle suspension, this geometry is considered at present the best for three phase suspension. Though it is a relatively inefficient geometry at low gassing rates (see Fig. 8), it becomes more efficient than the downward pumping impellers at high ones and it requires the least power of all to cause gas dispersion.

Again, N_{JSg} is greater than N_{CD} except for low density particles. Chapman et al.[53] confirmed that N_{CD} could be determined by the minimum in the graphs of Po_g vs. Fl_G (Fig. 10) and predicted quite accurately in vessels from 0.3 to 1.8 m in diameter by the equation[57]

$$(Fl_G)_{CD}^{-0.5}(Fr)_{CD}^{0.25}(D/T)^{0.25} = 2.25 \qquad (24)$$

or

$$N_{CD}D^2/Q_G^{0.5}T^{0.25} = 4 \text{ m}^{0.25}\text{ s}^{-0.5}. \qquad (25)$$

In addition, ΔN_{JS} for $C = T/4$ and $D = T/2$ for a wide range of particle sizes and density is found to increase linearly with gassing rate (expressed as vvm) in vessels from 0.29 to 1.8 m diameter (Fig. 11).

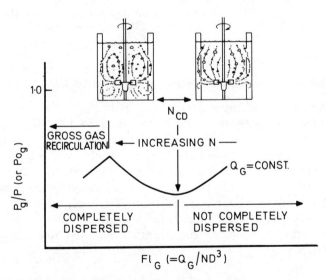

Figure 10   Idealised link between observed N_{CD} and its determination from power measurements for disc turbines[57].

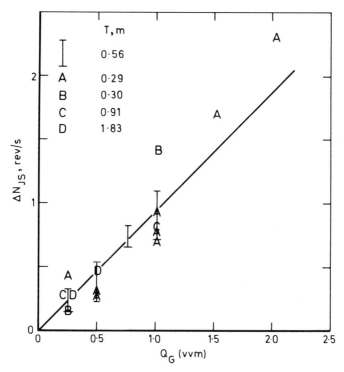

Figure 11 ΔN_{JS} vs. vvm for 5 vessel sizes for 6-bladed disc turbines ($H = T/2$; $C = H/4$; $1\% \leqslant X \leqslant 30\%$; $1050 \text{ kg/m}^3 \leqslant \rho_s \leqslant 2900 \text{ kg/m}^3$; $90 \,\mu\text{m} \leqslant d_p \leqslant 2650 \,\mu\text{m}$; in water at $\sim25°\text{C}$)[43].

6.5.   Effect of Physical Properties and Concentration

Chapman et al.[43] used particles from 1050 to 2900 kg/m^3 and of size from 80 μm to 2800 μm at concentration up to 30% by weight in liquids from 1 to 5 m Pas. and Fig. 11 is based on this range of conditions. The work was conducted in vessels from 0.29 to 1.8 m diameter with a range of impeller types (see Fig. 8). N_{JSg} varied with d_p, X, and ν approximately as it does in ungassed systems and the use of a mass mean for a distribution of particle sizes was again found to be applicable. Though still the most important property variable, the dependence on density difference was somewhat reduced, that is,

$$N_{JSg} \propto \Delta \rho^{0.22} \tag{26}$$

6.6. A Tentative Design Method for Disc Turbines

Sufficient information is available to determine the agitation speed and power input for three phase systems. Firstly, a disc turbine of diameter $T/2$ and of clearance $H/4$ is chosen as it is energy efficient for both gas dispersion and particle suspension and does not exhibit instabilities. N_{JS} can then be calculated from Eq. (14) and Fig. 4. ΔN_{JS} can then be determined from Fig. 11 to give N_{JSg}. N_{CD} should also be calculated from Eq. (24) and it should then be confirmed that for the gassing rate of interest, N_{CD} is less than N_{JSg}. For the determination of the power required, reference should be made to the extensive work on gassed agitation (Chapter 6, Nienow et al.[57] and Chapman et al.[53]). It should however be noted that the presence of the solids has a negligible effect on the power drawn by the impeller under gassed conditions up to about 30% by weight[43].

7. PARTICLE-TO-LIQUID TRANSFER PROCESSES

7.1. The Transfer Rate and the State of Suspension

The transfer of heat or mass between a particulate solid and a liquid is related to the physical properties of the fluid, the hydrodynamics in the vicinity of the particle and the interfacial area available. For transfer processes in agitated vessels, experimentally determined coefficients are normally calculated assuming all of the surface area of the particles is available. In fact, this is not the case for agitator speeds less than N_{JS}. The assumption is made because of the difficulty of determining precisely the actual area at agitator speeds less than N_{JS}. This assumption has two consequences. Firstly, the transfer coefficient increases relatively rapidly for agitator speeds less than N_{JS} as the area available increases and the resistance to mass transfer in the fluid phase decreases; and for speeds greater than N_{JS}, the rate increases relatively slowly as only the second of these two effects is involved. Secondly, any theoretically-based equation for calculating transfer rates can only be developed for agitator speeds of N_{JS} and above, that is, for agitation conditions where the surface area of contact is known.

7.2. Heat and Mass Transfer Equations

Of the many equations which might be used, a form of the Froessling equation for mass transfer between a sphere and a fluid

$$Sh = 2 + 0.73 \, Re_p^{1/2} \, Sc^{1/3} \qquad (27)$$

has been used successfully by a number of workers[22, 31, 58, 59]. The constant 0.73 is used here following the extensive analysis of literature data by Rowe et al.[60] and the equivalent equation for heat transfer proposed by the same workers should be applicable too.

At first, this choice of equation appears rather surprising as the Froessling equation is derived from an analysis which assumes steady flow over a sphere forming a laminar boundary layer. However, it has also been used successfully in the case of liquid-fluidised spheres[61], and with steady turbulent flow[62] with only a slight modification. Changing to any other general mass transfer equation would basically only mean changing either the constants 2 or 0.73 or the Reynolds or Schmidt group exponent. The best equation can only be determined with any degree of accuracy by using very precise data and even then the best may not be significantly better statistically than the Froessling equation[60].

The exponent on the Schmidt group is probably as good a guide as any as to which should be used because it does not require any assumptions about the velocity term in the Reynolds group. Though many early papers on stirred systems used $(Sc)^{1/2}$, the exponent was first chosen because it had been used successfully in absorption studies and other workers just followed this lead. Also, early diffusivity data were notoriously unreliable. Recent workers with accurate data have found $(Sc)^{1/3}$ correlates their mass transfer data better. This is another justification for the use of the Froessling equation.

7.3. Particle-Fluid Slip Velocity

In order to use Eq. (27) or the equivalent heat transfer equation, the particle Reynolds number must be known. This is defined as $Re_p = u_s d_p / \nu$ where u_s is the particle-fluid slip velocity, that is, it is a measure of the relative velocity between the particle and the liquid. The slip velocity in the extremely complex, three-dimensional turbulent flow occurring in a stirred vessel is difficult to define precisely[5]. Particles can move at a different gross velocity to the fluid in both magnitude and direction (Fig. 12a). However, in addition, particles can rotate relative to the fluid (Fig. 12b). The fluid's small-scale turbulent eddies may be much smaller than the particle, giving a relative velocity at the surface even though the particle is in general moving with the fluid (Fig. 12c). In the most complex and

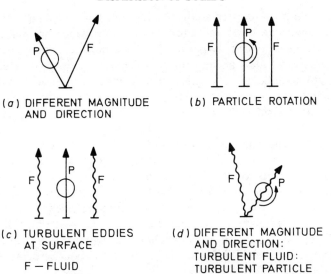

(a) DIFFERENT MAGNITUDE
AND DIRECTION

(b) PARTICLE ROTATION

(c) TURBULENT EDDIES
AT SURFACE

F — FLUID
P — PARTICLE

(d) DIFFERENT MAGNITUDE
AND DIRECTION:
TURBULENT FLUID:
TURBULENT PARTICLE
MOTION AND ROTATION:

Figure 12 Conceptual slip velocities[5].

general real case, the particles will themselves move turbulently, that is, with a mean velocity and direction at any point but with random fluctuations in any direction, and they will rotate. The fluid will also be turbulent and will have a mean velocity different in magnitude and direction from the fluid (Fig. 12d). Finally, both particle and fluid velocities will vary spatially throughout the vessel.

Measurement of this complex slip velocity in specific regions of the vessel has proved unsuccessful[5, 21, 22], nor is it obvious if it were measured how the correct average value would be obtained. Many attempts to estimate it have been made invoking Kolmogoroff's theory of isotropic turbulence. However, strictly this theory only gives directly estimates of eddy sizes and eddy velocities. Assumptions must be made in order to link eddy size and velocity to the particle size and the slip velocity. For example, Eq. (6) for eddy velocities in the inertial subrange is converted to slip velocities for particles of size d_p by assuming, as for Eq. (6), that d_p equals λ and, in addition, u_s is proportional to u', that is,

$$u_s \propto (d_p \epsilon_T)^{1/3} \tag{28}$$

The other assumption is that though the energy dissipation rate varies from place to place, substitution of the average rate is satisfactory so that Re_p is given by $(d_p^{4/3} \epsilon_T^{1/3}/\nu)$.

However, as indicated in the section on Fluid Mechanics, it is the large convective flows and anisotropic eddies which are paramount in particle suspension. Experiments show that the same mass transfer rates are found at very different energy dissipation rates with different impellers[20, 63]. It must be concluded, therefore, that conceptually and experimentally, there are anomalies in the application of Kolmorgoroff's theory to particle–fluid transfer processes in stirred vessels[64].

Another approach, averaging and simplifying the complex reality of the problem, is to link (equate) the slip velocity at N_{JS} to the terminal velocity of the particle in a still fluid, u_T. Since N_{JS} is dependent on the bulk convective flows and anistropic eddies, this approach links these bulk flows to mass transfer rates. This assumption is obviously a gross one, but it does imply that the mass transfer coefficient at N_{JS} should be independent of agitator type and energy disipation rate. This is found to be in quite good agreement with experiments[20, 31, 65].

In fact, the measured mass transfer coefficient is greater than that predicted by the subsitution of u_T into Eq. (27). It is possible to correlate this enhancement either with particle size[66] or with the Kolmogoroff-based Re_p[59]. It has been proposed that the reason for the enhancement is the increase in u_s above u_T due to agitation affecting the mechanisms depicted in Fig. 12.

For agitator speeds greater than N_{JS}, the mass transfer coefficient increases at a rate dependent on particle size, particle density and system geometry[31, 67]. This implies that the slip velocity is again dependent on the large convective flows and anisotropic eddies (affected by system geometry) and the spacial distribution of the particles (dependent on their size and density) amongst them. Thus, these interactions are extremely complex and system dependent. Because of this complexity, no entirely satisfactory method of determining the mass transfer coefficient at agitator speeds greater than N_{JS} is available. However, a simplified analysis has been proposed to enable u_s to be estimated for this region[68].

7.4. The Calculation of Mass Transfer Coefficients

The mass transfer coefficient is sensitive to system geometry and to the particle/fluid properties[63]. Because of this complexity, the following simplified procedure is recommended. It is based on the premise that the geometry which requires the least energy dissipation for particle suspension will also be the most energy efficient for mass transfer. The procedure is:

1. Select a system configuration which requires low energy dissipation rates for particle suspension. This can also be done for three phase systems though gas–liquid dispersion and mass transfer rates also need to be considered (Chapman et al.[8, 43, 53], and Chapter 6).
2. Calculate N_{JS} as set out in this chapter.
3. Calculate the terminal velocity of the particle[31] recognising that, for large particles, an enhanced drag coefficient ($C_D = 1.0$) must be employed due to the high level of turbulence in the fluid[5, 21].
4. Determine the mass transfer coefficient for this velocity from Eq. (27).
5. Allow for enhancement above this value due to system turbulence[59, 66].
6. Because of the complexity of the relationships between the mass transfer coefficient and the agitation conditions at agitator speeds greater than N_{JS}, reference should be made to the original papers[31, 59, 66]. However, it is important to note that for geometrical configurations which require low energy dissipation rates for particle suspension, the mass transfer coefficient increases only as $N^{0.5}$ whilst ϵ_T increases as N^3. Therefore a speed greater than N_{JS} is often not justified.

8. CONCLUSIONS

No entirely satisfactory models for particle suspension in stirred vessels are presently available. This is because of the complexity of the three-dimensional flow within them. Processes which are particularly dependent on the small scale turbulence structure are often capable of satisfactory modelling by the application of Kolmogoroff's theory of isotropic turbulence. However, particle suspension is dependent on the bulk convective flows and the large scale, anisotropic eddies especially those near the base. Until some means of linking these large scale flow structures to suspension mechanisms is found, (by applying computer simulation techniques, for example), good experimental measurements and accurate empirical correlations guided by some simple theoretical reasoning will be the best that can be achieved.

The just completely suspended condition is the most important one in solids dispersion. In ungassed systems even after a quarter century, the correlation first proposed by Zwietering is the best that is available. It can be extended to cover size distributions by the use of a mass mean diameter. For scale-up, because of the sensitivity of power to experimental errors and parameters determined from experimental data, it is recommended that the rule

$$(\epsilon_T)_{JS} \propto D^{-0.28} \tag{29}$$

should be used following the application of Zwietering's equation [Eq. (14)] to calculate N_{JS} on the small scale. Downward pumping axial flow impellers at a clearance above the base of about 1/5 require the least energy and the use of a draught tube and a contoured-bottom can greatly reduce this.

For continuous withdrawal of solids, their residence time distribution is as much affected by the mode of withdrawal as by the state of suspension. Means of allowing for deviations for C.S.T.R. behaviour are available.

Little has been done on dispersing floating solids and more work is required.

For solid dispersion in gassed systems, downward pumping impellers require least energy at low gas flow rates but are subject to gross flow instabilities and are very sensitive to gassing rate. Upward pumping impellers require most energy at low gas rates but little extra energy is required up to about 1 vvm. Therefore, at high gas rates, they are the most efficient. No instabilities are found.

Disc turbines exhibit instabilities at low clearance ($C/T = 1/6$) but not at $C/T = 1/4$. Large impellers ($D = T/2$) require least energy for both gas dispersion and particle suspension. Both the speed to completely disperse the gas and just suspend solids can be calculated from equations available in the literature and a method of determining the increase in speed required to keep solids in suspension, due to the introduction of gas, is given. Taking into account the availability of this design procedure, the large body of literature on its gassed power consumption, its stability over all gassing rates up to 1 vvm in vessels up to 1.8 m diameter, the $T/2$ disc turbine at a clearance of $H/4$ seems to offer the safest impeller type for three phase agitation at present.

Because of its lower power requirements when suspending solids at gas rates $\geqslant 1$ vvm, the upward pumping impeller may have very distinct advantages at very high rates and further work on it would be worthwhile.

Transfer rates to and from particles are best estimated from a Froessling-type equation in which the particle–fluid slip velocity is estimated from the still-fluid terminal velocity. The actual transfer rate is greater than this due to the turbulence level in agitated vessels.

9. NOTATION

A constant in Eq. (1)

a exponent

b exponent

B baffle width, m

C agitator clearance above base, m

C_{out} solids concentration in outlet flow, kg/m^3

$\bar{\bar{C}}$ solids concentration in vessel, kg/m^3

C_D drag coefficient, dimensionless

D agitator diameter, m

$\mathscr{D}$ mass diffusion coefficient, m^2/s

d_p particle size, m

Fl_G gas flow number (Q_G/ND^3), dimensionless

Fr Froude number, dimensionless

g gravitational constant, $9.81 \ m/s^2$

H liquid height in vessel, m

K steady state separation coefficient, dimensionless

k mass transfer coefficient, m/s

N agitator speed, rev/s

N_{PJS} agitator speed at which an instrument indicates complete suspension, rev/s

ΔN_{JS} $N_{JSg} - N_{JS}$ rev/s

P power input to the liquid, W

Po power number, dimensionless

Q_G sparged gas flow rate, m^3/s (or vvm if stated)

Re Reynolds number for agitation, dimensionless

Re^* Reynolds number for suspension, dimensionless

Re_K Reynolds number of Kolmogoroff eddy ($=1$)

Re_P particle Reynolds number, $u_s d_p/\nu$, dimensionless

S Zweitering's suspension parameter [see (Eq. (14)], dimensionless

Sc Schmidt number ($\nu/\mathscr{D}$), dimensionless

Sh Sherwood number, $kd_p/\mathscr{D}$, dimensionless

T vessel diameter, m

t time, s

$\bar{u}$ the mean velocity (modulus) at a level in the vessel, m/s

u' fluctuating velocity of an eddy, m/s

u_s particle–fluid slip velocity (see Fig. 12), m/s

u_T particle terminal velocity in a still fluid, m/s

V vessel volume, m^3

X (wt of solids in the suspension per weight of liquid $\times$ 100), dimensionless

Z a suspension parameter [see Eq. (8)], dimensionless

Greek Symbols

ϵ_T mean energy dissipation rate throughout the vessel or specific power input, W/kg

λ the size of a turbulent eddy, m

ν kinematic viscosity, m^2/s

ρ_L fluid density, kg/m^3

ρ_s solid density, kg/m^3

$\Delta\rho$ $\rho_s - \rho_L$, kg/m^3

τ_L liquid mean residence time, s

τ_S solid mean residence time, s

Subscripts

base at the base of the vessel

CD the condition of agitation which just completely disperses the sparged gas

DF the condition of agitation at which floating particles are all just dispersed into the fluid

g under conditions of gas sparging

JH the condition of agitation at which particles just become homegeneously suspended.

JS the condition of agitation at which particles just become completely suspended based on visual observation

K at the Kolmogoroff eddy size

90 the condition of agitation which causes suspended solids to reach to 90% of the liquid height

REFERENCES

1. Parfitt, G. D. (1973). *Dispersion of Powders in Liquids*, 2nd Edn., App. Sci. Publ., London.
2. Nienow, A. W., and Conti, R. (1978). Particle abrasion at high solids concentration in stirred vessels, *Chem. Eng. Sci.*, **33**, 1077.

3. Nienow, A. W. (1976). The effect of agitation and scale-up on crystal growth rates and secondary nucleation, *Trans. Inst. Chem. Eng.*, **54**, 205.

4. Cutter, L. A. (1966). Flow and turbulence in a stirred tank, *AIChE J.*, **12**, 35.

5. Nienow, A. W., and Bartlett, R. (1975). The measurement and prediction of particle-fluid slip velocities in agitated vessels, *Proc. 1st European Conf. on Mixing and Centrifugal Separation*; BHRA, Cranfield, Bedford, England, pp. B1–15.

6. Bourne, J. R., and Zabelka, M. (1980). The influence of gradual classification on continuous crystallisation, *Chem. Eng. Sci.*, **35**, 533.

7. Nienow, A. W. (1968). Suspension of solid particles in turbine-agitated, baffled vessels, *Chem. Eng. Sci.*, **23**, 1453.

8. Chapman, C. M., Nienow, A. W., Cooke, M., and Middleton, J. C. (1983) Particle–gas–liquid mixing in stirred vessels: part 1, Particle–liquid mixing, *Trans. Inst. Chem. Eng. (Chem. Eng. Res. Des.)*, **61**, 71.

9. Chapman, C. M. (1981). Studies of gas–liquid–particle mixing in stirred vessels, Ph.D. Thesis, University of London.

10. Cleaver, J. W., and Yates, B. (1973). Mechanisms of detachment of colloidal particles from a flat substrate in turbulent flow, *J. Coll. Int. Sci.*, **44**, 464.

11. Ohiaeri, I. (1981). The suspension of solids in mixing vessels, Ph.D. Thesis, University of Bradford.

12. Levich, V. G. (1962). *Physico-Chemical Hydrodynamics*, Prentice-Hall, Englewood Cliffs, New Jersey.

13. Quraishi, A. Q., Mashelkar, R. A., and Ulbrecht, J. J. (1977). Influence of drag reducing additives on mixing and dispersion in agitated vessels, *AIChE J.*, **23**, 487.

14. Aesbach, S., and Bourne, J. R. (1972). Attainment of homogeneous suspensions in a continuous stirred tank, *Chem. Eng. J.*, **4**, 234.

15. Einenkel, W. D., and Mersmann, A. (1977). The agitator speed for particle suspension, *Verfahrenstechnik*, **11**, 90, (in German).

16. Nienow, A. W., and Wisdom, D. J. (1974). Flow over disc turbine blades, *Chem. Eng. Sci.*, **29**, 1994.

17. Van't Riet, K., and Smith, J. M. (1975). The trailing vortex system produced by Rushton turbine agitators, *Chem. Eng. Sci.*, **30**, 1093.

18. Tatterson, G. B., Hsien Kwa, S. Y., and Brodkey, R. S. (1980). Stereoscopic visualisation of the flows for pitched blade turbines, *Chem. Eng. Sci.*, **35**, 1369.

19. Bruin, W., van't Riet, K., and Smith, J. M. (1974). Power consumption with aerated Rushton turbines, *Trans. Inst. Chem. Eng.*, **52**, 88.

20. Nienow, A. W., and Miles, D. (1978). The effect of impeller/tank configurations on fluid–particle mass transfer, *Chem. Eng. J.*, **15**, 13.

21. Schwartzberg, H. G., and Treybal, R. E. (1968). Fluid and particle motion in turbulent stirred tanks, *Ind. Eng. Chem. Fund.*, **7**, 1.

22. Levins, D. M., and Glastonbury, J. R. (1972). Particle–liquid hydrodynamics and mass transfer in a stirred vessel, *Trans. Inst. Chem. Eng.*, **50**, 32 and 132.

23. Bates, R. L., Fondy, P.L., and Corpstein, R. R. (1963). An examination of some geometric parameters on impeller power, *Ind. Eng. Chem. Proc. Des. Dev.*, **2**, 310.

24. Zwietering, T. N. (1958). Suspending solid particles in liquids by agitators, *Chem. Eng. Sci.*, **8**, 244.

25. Gates, L. E., Morton, J. R., and Fondy, P. L. (1976). Selecting agitator systems to suspend solids in liquids, *Chem. Eng.*, **83** (Apr. 26th), 102.

26. Baldi, G., Conti, R., and Alaria, E. (1978). Complete suspension of particles in mechanically agitated vessels, *Chem. Eng. Sci.*, **33**, 21.

27. Oroskar, A. R., and Turian, R. M. (1980). The critical velocity in pipeline flow of slurries, *AIChE J.*, **26**, 550.

28. Middleton, J. C. (1983). I.C.I. Heavy chemicals new science group, private communication.

29. Fort, I. (1983). Prague Institute of Chemical Technology, Private communication.
30. Hixson, A. W., and Baum, S. J. (1941). Mass transfer coefficients in liquid–solid agitation systems, *Ind. Eng. Chem.*, **33**, 478.
31. Nienow, A. W. (1969). Dissolution mass transfer in a turbine agitated baffled vessel, *Can. J. Chem. Eng.*, **47**, 248.
32. Nienow, A. W. (1981). Suspension of solids in liquids, I. Chem. E. Post experience course on "Mixing in the Process Industries", University of Bradford, 20 pp.
33. Nienow, A. W. (1981). The mixer as a reactor: liquid–solid systems, I. Chem. E. Post Experience Course on "Mixing in the Process Industries", University of Bradford, 18 pp.
34. Nienow, A. W., and Miles, D. (1971). Impeller power numbers in closed vessels, *Ind. Eng. Chem. Proc. Des. Dev.*, **10**, 41.
35. Bourne, J. R., and Sharma, R. N. (1974). Homogeneous particle suspension in propeller-agitated flat bottomed tanks, *Chem. Eng. J.*, **8**, 243.
36. Musil, L. (1976). The hydrodynamics of mixed crystallisers, *Coll. Czech. Chem. Commun.*, **41**, 839.
37. Conti, R., and Baldi, G. (1978). Continuous removal of a solid suspension from stirred tanks, *Proc. Int. Symp. on Mixing*, Mons, Belgium, Paper B2.
38. Machon, V., Kurdna, V., and Hudcova, V. (1980). The steady state separation coefficient for solids flowing through an agitated vessel, *Coll. Czech. Chem. Commun.*, **45**, 2152.
39. Mattern, R. U., Bilous, O., and Piret, E. J. (1957). Continuous-flow stirred tank reactors: solid–liquid systems, *AIChE J.*, **3**, 497.
40. Zacek, S., Nyvelt, J., Garside, J., and Nienow, A. W. (1982). A stirred tank for continuous crystallisation, *Chem. Eng. J.*, **23**, 111.
41. Jahnse, A. K., and De Jong, E. J. (1977). The importance of classification in well-mixed crystallisers, in *Industrial Crystallization* (Ed. J. W. Mullin), Plenum Press, New York, p. 403.
42. Mersmann, A., Einenkel, W. D., and Kappel, M. (1976). Design and scale-up of mixing equipment, *Int. Chem. Eng.* **16**, 590.
43. Chapman, C. M., Nienow, A. W., Cooke, M., and Middleton, J. C. (1983). Particle–gas–liquid mixing in stirred vessels: part 3, Three–phase mixing, *Trans. Inst. Chem. Eng. (Chem. Eng. Res. Des.)*, **61**, 167.
44. Weisman, J., and Efferding, L. E. (1960). Suspension of slurries by mechanical mixers, *AIChE J.*, **6**, 419.
45. Rieger, F., Ditl, P., and Novak, V. (1978). Suspension of solid particles in agitated vessels, *CHISA Congress*, Prague, Paper A5.3.
46. Cliff, M. H., Edwards, M. F., and Ohiaeri, I. (1981). The suspension of settling solids in agitated vessels, *Proc. Conf. on Fluid Mixing*, I. Chem. E. Symp. Ser. No. 64, pp. M1–11.
47. Joosten, G. E. H., Schilder, J. G. M., and Broore, A. M. (1977). The suspension of floating solids in stirred vessels, *Trans. Inst. Chem. Eng.*, **55**, 220.
48. Chaudhari, R. V., and Ramachandran, P. A. (1980). Three phase slurry reactors, *AIChE J.*, **26**, 179.
49. Arbiter, N., Harris, C., and Yap, R. F. (1969). Hydrodynamics of flotation cells, *Trans. A.I.M.E.*, **244**, 134.
50. Weidmann, J. A., Steiff, A., and Weinsbach, P. M. (1980). Experimental investigations of suspension, dispersion, power, gas hold-up and flooding characteristics in stirred gas–liquid–solid systems (slurry reactors), *Chem. Eng. Commun.*, **6**, 245.
51. Subbarao, D., and Taneja, V. K. (1979). Three phase suspensions in stirred vessels, *Proc. 3rd European Conf. on Mixing*, B.H.R.A., Cranfield, Bedford, England, p. 229.
52. Sicardi, S., Conti, R., Baldi, G., and Franzino, L. (1980). Suspension of particles and solid–liquid mass transfer in slurry stirred reactors, *Proc. 2nd Yugoslav–Italian–Austrian Chem. Eng. Conf.*, Bled. p. 452.

53. Chapman, C. M., Nienow, A. W., Cooke, M., and Middleton, J. C. (1983). Particle–gas–liquid mixing in stirred vessels: part 2, Gas–liquid mixing, *Trans. Inst. Chem. Eng. (Chem. Eng. Res. Des.)*, **61**, 82.
54. Kuboi, R., and Nienow, A. W. (1982). The power drawn by dual impeller systems under gassed and ungassed conditions, *Proc. 4th European Conf. on Mixing*, BHRA, Cranfield, Bedford, England, pp. 247–261.
55. Nienow, A. W., Kuboi, R., Chapman, C. M., and Allsford, K. (1983). The dispersion of gases into liquids by mixed flow agitators, *Proc. Int. Conf. on Physical Modelling of Multiphase Flows*, BHRA, Cranfield, Bedford, England, pp. 417–438.
56. Chapman, C. M., Nienow, A. W., and Middleton, J. C. (1981). Particle suspension in a gas-sparged, Rushton turbine-agitated vessel, *Trans. Inst. Chem. Eng.*, **59**, 134.
57. Nienow, A. W., Wisdom, D. J., and Middleton, J. C. (1978). The effect of scale and geometry on flooding, recirculation and power in gassed, stirred vessels, *Proc. 2nd European Conf. on Mixing*, B.H.R.A., Cranfield, Bedford, England, pp. F1–1 to F1–16.
58. Harriott, P. (1962). Mass transfer to particles suspended in agitated tanks, *AIChE J.*, **8**, 93.
59. Conti, R., and Sicardi, S. (1982). Mass transfer from freely-suspended particles in stirred tanks, *Chem. Eng. Commun.*, **14**, 91.
60. Rowe, P. N., Claxton, K. T., and Lewis, J. B. (1965). Heat and mass transfer from a single sphere in an extensive flowing fluid, *Trans. Inst. Chem. Eng.*, **43**, T14.
61. Rowe, P. N., and Claxton, K. T. (1965). Heat and mass transfer from a single sphere to a fluid flowing through an array, *Trans. Inst. Chem. Eng.*, **43**, T321.
62. Galloway, T. R., and Sage, B. H. (1964). Thermal and material transfer in turbulent gas streams, *Int. J. Heat Mass Trans.*, **7**, 283.
63. Aussenac, D., Alran, C., and Couderc, J. P. (1982). Mass transfer between suspended solid particles and a liquid in a stirred vessel, *Proc. 4th European Conf. on Mixing*, BHRA, Cranfield, Bedford, England, pp. 417–421.
64. Levins, D. M., and Glastonbury, J. R. (1972). Application of Kolmogoroffs theory to mass transfer in agitated vessels, *Chem. Eng. Sci.*, **27**, 537.
65. Ovsenik, A. (1982). Optimisation of solid particles suspending, *Proc. 4th European Conf. on Mixing*, BHRA, Cranfield, Bedford, England, pp. 463–470.
66. Nienow, A. W. (1975). Agitated vessel particle–liquid mass transfer: a comparison between theories and data, *Chem. Eng. J.*, **9**, 153.
67. Nagata, S. (1975). *Mixing*, Kondansha, Tokyo.
68. Nienow, A. W., Bujac, P. D. B., and Mullin, J. W. (1972). Slip velocities in agitated vessel crystallisers, *J. Cryst. Growth*, **13/14**, 488.

Chapter 9

Current Trends in Mixer Scale-up Techniques

JAMES Y. OLDSHUE

Mixing Equipment Company, 135 Mt. Read Blvd., Rochester, NY 14611, USA

1. INTRODUCTION

Scale-up involves the translation of data on a small scale to production size equipment. Mixing involves many different kinds of operations, and there is no one scale-up rule that applies to all of these
operations. This article summarizes the difference in many physical
and mass transfer mixing parameters between large tanks and small
tanks, and reviews some of the rationale for using different scale-up
concepts for various kinds of mixing operations.

In particular, a look will be taken at geometric similarity, dynamic
similarity, and dimensionless groups, since they are basic tools.
Their applicability is a key factor in this review of current practice.

Table 1 shows some overall concepts. Given a geometry, process
design involves choosing any two of the three variables of power, P,
speed, N, and diameter, D. In the process area, there are correlations which describe the mixer's process performance. Such
correlations involve fluid properties, tank geometry, and only two of
the three variables, P, N, or D because the third variable is fixed for
us by the power correlation (Fig. 2), which is independent of process
performance. The third variable can be calculated, but it does not
enter into the process correlation. I prefer to use power and
diameter as the two parameters in the process correlation. It does
not matter mathematically if you use speed and diameter or speed
and power in the process correlation.

Table 1   Elements of Mixer Design.

I. Process design	1. Fluid mechanics of impellers
	2. Fluid regime required by process
	3. Scale up; hydraulic similarity
II. Impeller power characteristics	4. Relates impeller h.p., speed & diameter
III. Mechanical design	5. Impellers
	6. Shafts
	7. Drive assembly

When one looks at the fluid flow pattern required by a process, one must look at the mechanical combinations involved and pick out the combination that has the optimum cost for the application. There is normally no one unique mixer for a given application. There are several, and the best selection is the one which is optimum in terms of overall economics of the system.

Table 2 lists five basic application classes. The most important distinction is to divide them into the concept of physical uniformity, in the left column, in contrast to the concept of mass transfer, chemical reaction, or some sort of diffusion processes in the right. The things done for effective solids suspension can be quite different than those done for solids dissolving. The things done to disperse a gas are quite different from what is done to carry out a gas adsorption. There are ten separate mixing technologies. Each of these ten areas has its own scale-up rules and principles, application know-how, application experience, etc. Let us look at some examples and see how one prepares a pilot plant program. The overall process result may be made up of many component steps. Our job is to organize it in a way that makes those steps meaningful and yet manageable in terms of present day data.

First, scale-up principles will be discussed. Then, secondly, let us look at how to conduct specific experiments to determine the controlling factors for scale-up of the process. There is no shortage of scale-up principles or possibilities. The right one must be applied to a given circumstance.

Table 2   Different Types of Mixing Processes.

Physical processing	Application classes	Chemical processing
Suspension	Liquid–Solid	Dissolving
Dispersions	Liquid–Gas	Absorption
Emulsions	Immiscible liquids	Extraction
Blending	Miscible liquids	Reactions
Pumping	Fluid Motion	Heat Transfers

In order to study a fluid regime for mixing, a process result must be in mind that we are trying to achieve. A flow pattern cannot be determined to be good or bad by looking at it, unless there is an objective from a process standpoint.

Even then, an evaluation cannot be made unless we have an installation and economics in mind. Is a 50 gallon tank involved, in which power is a very minor consideration, or a 50,000 gallon tank where power and other costs are important.

The economic basis to be used to evaluate the mixer variables must be available. For example, in the waste treating industry in the USA, the cost of power over 20 years leads to evaluation of mixers at roughly $5000 per HP. $5000 may be spent now to save one HP in the future over 20 years. In a chemical plant where it is usual to use one to three years for evaluation, the costs are about $500 to $1000 per HP saved. The effort required to optimize mixer variables in the laboratory depends heavily on the economic basis.

If a pilot plant process was successful, and it was deemed to duplicate every mixing parameter in that pilot plant tank, a large number of tanks exactly the same size would be needed, because once there is a bigger vessel, the parameters are different. The bigger vessel normally has a longer blend time, it has a higher maximum shear rate, a lower average shear rate, and a greater variety of shear rates[1].

Fortunately, many processes are not that sensitive to all these differences, and the geometrically similar full-scale tank behaves successfully. But when one looks in detail at what happens at the molecular level, or to the fluid shear rates or pumping capacities, it is found that the big tank is quite different from the small tank. That gives the challenge of deciding which differences are of major importance and which are of minor importance.

In order to indicate how things change on scale-up, Table 3 is a chart on which a few of the parameters are shown, such as power, power per unit volume, speed, diameter, pumping capacity of the impeller, pumping capacity of the impeller per unit volume, tip speed, and Reynolds number.

A value of one is assigned to each variable just to see how they change relative to each other on scale-up. There are four different scale-up calculations; in column 3, power per unit volume is held constant but everything else changes. Speed goes down, pumping capacity per unit volume goes down, while the tip speed and Reynolds number go up. In column 4, there is constant pumping capacity per unit volume. Power per unit volume increases with the

Table 3   Properties of a Fluid Mixer on Scale-Up.

PROPERTY	PILOT SCALE 20 GALLONS	PLANT SCALE 2500 GALLONS			
P	1.0	125	3125	25	0.2
P/VOL.	1.0	1.0	25	0.2	0.0016
N	1.0	0.34	1.0	0.2	0.04
D	1.0	5.0	5.0	5.0	5.0
Q	1.0	42.5	125	25	5.0
Q/VOL.	1.0	0.34	1.0	0.2	0.04
ND	1.0	1.7	5.0	1.0	0.2
$\dfrac{ND^2\rho}{\mu}$	1.0	8.5	25.0	5.0	1.0

square of the tank diameter. Trying to maintain equal circulation time is usually impractical on a full-scale tank.

In column 5, at equal tip speed, the power per unit volume has actually dropped on scale-up inversely with scale size ratio. In column 6, Reynolds number is constant, and that decreases the total power. In general, the Reynolds number will go up, circulation times will go up, and most of the time, the tip speed of the impeller will go up on scale-up. It is difficult to control all the ratios we may be interested in.

Is it possible that for every mixing process there is a constant scale-up parameter somewhere? Trying to force each application into a constant parameter normally does not work. If one thinks of these as correlating parameters, and then finds out how they change on scale-up, everything will work out better. When a parameter is a constant, it simplifies the mathematics.

In a fluid mixing application there are four fluid forces that are important (see Table 4). There is the inertia force that is put in by the mixer, and there are opposing forces of viscosity, gravity, and surface tension. Ideally, we would like all the force ratios to be constant between the model and the prototype. The problem is that in a mixing vessel with the same fluid in both small and large tanks, all these ratios cannot be constant. Only two of the four can be kept constant. These ratios relate to the flow pattern in the vessel, but not to the complexity of molecular diffusion and mass transfer. Dynamic similarity is only one measure of similarity, but not the whole story of mixing similarity.

Table 4   Pertinent Ratios in Hydraulic Similitude and in particular Dynamic Similitude. Subscript "M" is Model, and subscript "P" is Prototype.

Geometric $\dfrac{X_M}{X_P} = X_R$

Dynamic $\dfrac{(F_I)_M}{(F_I)_P} = \dfrac{(F_V)_M}{(F_V)_P} = \dfrac{(F_G)_M}{(F_G)_P} = \dfrac{(F_\sigma)_M}{(F_\sigma)_P} = F_R$

Table 5   Force Ratios, including the Reynolds number, the Froude number, and the Weber number.

$\dfrac{F_1}{F_v} = N_{Re} = \dfrac{ND^2\rho}{\mu}$

$\dfrac{F_1}{F_g} = N_{Fr} = \dfrac{N^2D}{g}$

$\dfrac{F_1}{F_\sigma} = N_{We} = \dfrac{N^2D^3\rho}{\sigma}$

The use of dimensionless ratios becomes very appropriate on occasion (see Table 5). The Reynolds number is the ratio of inertia force to viscous force, the Froude number is the ratio of inertia force to gravitational force, and the Weber number is the ratio of inertia force to surface tension. These are the ratios of what we put in the tank, the inertia force divided by either the opposing force of viscosity, or gravity, or surface tension. These ratios do work very well in certain cases as indicated through examples below.

1.1.   Power Correlation Example

In looking at the power drawn by a mixer, the Reynolds number–power number curve, Fig. 2, was developed by observing that power was a function of speed, density, viscosity, gravity, tank diameter, off-bottom distance and baffle width (Fig. 1). The applied force divided by the fluid acceleration, F/ma, is the power number for the mixing tank. In baffled tanks, the Froude number is not important, and the Reynolds number is all that is needed for a correlation (see Fig. 2). This is one of the best examples we have of the use of dynamic similarity and dimensionless groups in a mixing correlation.

IMPELLER   POWER   CHARACTERISTICS

$$P = f\left[D, N, \rho, \mu, g, T,\right]$$

$$\frac{APPLIED\ FORCE}{FLUID\ ACCELERATION} = f\left[\frac{APPLIED\ FORCE}{RESISTING\ FORCE}\right]$$

$$\frac{Pg}{\rho N^3 D^5} = \left[\frac{ND^2 \rho}{\mu}\right]^x \left[\frac{N^2 D}{g}\right]^y \left[\frac{D}{T}\right]^z$$

Figure 1   Variables involved in determining impeller power characteristics.

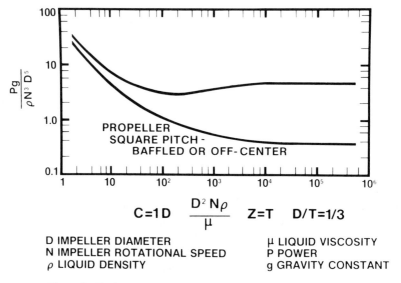

$$C = 1D \qquad \frac{D^2 N\rho}{\mu} \qquad Z = T \qquad D/T = 1/3$$

D IMPELLER DIAMETER   μ LIQUID VISCOSITY
N IMPELLER ROTATIONAL SPEED   P POWER
ρ LIQUID DENSITY   g GRAVITY CONSTANT

Figure 2   Typical correlation of Reynolds number, power number.

1.2. Heat Transfer Application

The heat transfer coefficient is a function of speed, diameter, density, viscosity, specific heat, thermal conductivity and tube size (see Fig. 3). A process group is composed of the process result, the heat transfer coefficient, divided by the system conductivity, and the tube diameter. This process group does correlate with the Reynolds number (see Fig. 4).

A word of caution: How can a dimensionless process group be written relating to such processes as polymerization, starch cooking, or crystallization? Those process results cannot be related to

APPLICATION OF HYDRAULIC SIMILARITY TO HEAT TRANSFER

$$h = f\left[N, D, \rho, \mu, Cp, k, d\right]$$

$$\frac{\textbf{PROCESS } \text{RESULT}}{\text{SYSTEM CONDUCTIVITY}} = f\left[\frac{\text{APPLIED FORCE}}{\text{RESISTING FORCE}}\right]$$

$$\frac{hd}{k} = \left[\frac{ND^2\rho}{\mu}\right]^x$$

Figure 3   Typical correlation variables for heat transfer process correlation.

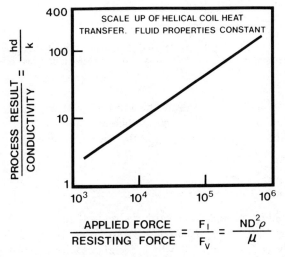

$$\frac{\text{APPLIED FORCE}}{\text{RESISTING FORCE}} = \frac{F_I}{F_V} = \frac{ND^2\rho}{\mu}$$

Figure 4   Typical correlation of the Nusselt number vs. Reynolds number for heat transfer.

dynamic force ratios. The bulk of mixing applications do not lend themselves to correlations of dimensionless process numbers with dimensionless fluid force numbers.

In the total heat transfer correlation[3], making use of all available dimensionless groups, such as the Prandtl number, the D/T ratio, etc., an equation results which correlates very well with Reynolds number (see Fig. 5).

With this data, it looks as if the flat blade turbine is a better heat transfer device than a propeller. What the curve does not show is that the power drawn by the turbine is 30 times the power drawn by the propeller at the same Reynolds number. If the speed of the

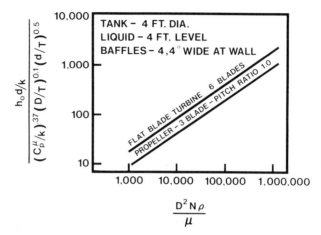

Figure 5 Complete correlation of Nusselt number, Prandtl number and other geometric ratios vs. Reynolds number for flat plate turbine and propeller.

propeller is increased to draw the same power, we find that we get about the same heat transfer coefficient. In actuality, at constant power, the radial flow turbine and the propeller give about the same heat transfer coefficients.

1.3.   Blending Process Example

Dimensionless groups also work quite well to correlate blending. The corresponding dimensionless group is the blend time times the impeller speed, (see Fig. 6), which correlates with Reynolds number. Different impellers give different correlation lines (see Fig. 7)[7]. The overall problem is that outside of these two or three examples, this dimensionless number technique cannot be applied to scale-up.

APPLICATION OF HYDRAULIC SIMILARITY TO BLENDING

$$\theta = f\left[N, D, \rho, \mu, T\right]$$

$$\frac{RESULT}{SYSTEM\ CONDUCTIVITY} = f\left[\frac{APPLIED\ FORCE}{RESISTING\ FORCE}\right]$$

$$\theta N \propto \left[\frac{ND^2\rho}{\mu}\right]^X \left[\frac{D}{T}\right]^Z$$

Figure 6 Typical correlating variables for blending scale-up process.

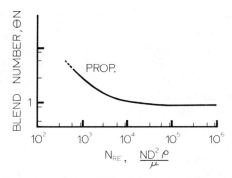

Figure 7   Typical correlation of Blend number and Reynolds number.

2.   PRACTICAL SCALE-UP CONSIDERATIONS

When the dimensionless group approach cannot be used, pilot plant runs must be made so it can be determined which ratios of which properties are important. Is it a mass transfer relationship, a shear rate relationship, or a pumping capacity relationship? When there is an idea of why the process behaves the way it does, then the ratios of those properties can be applied for scale-up.

Another characteristic about a mixing tank is the level of shear rates produced. As a calculation convenience, constant power per unit volume and geometric similarity is used in Fig. 8. The maximum shear rates around the impeller increase while the average shear

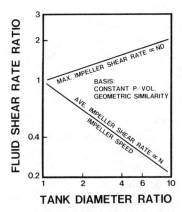

Figure 8   Indication that maximum impeller shear rate goes up while average impeller shear rate goes down for geometrically similar scale-up systems.

rates around the impeller decrease. One is related to the fact that tip speed has increased, and the other to the reduced operating speed. This shows that there is a greater range of shear rates in a big tank than in a small tank. It is very difficult to bring these curves together as we scale-up, but we can tip the whole ratio curve. To be successful, we have to use non-geometric similarity. Geometric similarity controls no mixing variable whatsoever. In order to scale-up mixing, the levels of the several most important variables must be established and, as a result, suffer the consequences of the variation of the other variables. All of the variables may not remain in the desired ratio, and geometric similarity may not be preserved, but two or three variables may be controlled to desired levels.

The major differences between a big tank and a small tank are that the big tank has a longer blend time, a higher maximum impeller shear rate, and a lower average impeller shear rate.

Looking at why a small tank does not model a big production tank, the small tank has too high a pumping capacity, and therefore too short a blend time. It has too low a maximum shear share rate to give the shear effects seen in the big tank. So in order to duplicate those two parameters, the impeller must be made narrower or smaller, or both, and be run faster, which cuts the pumping rate and increases the shear rate to be more like the plant mixer. That does not take care of all the other parameters[2]. If other variables have to be controlled, various pilot plant geometries or conditions may have to be used to test step by step the effects of the various parameters. Nongeometric similarity cannot include every important mixing variable. It can certainly make the pilot tank more similar to the large tank and many times enough similarities may be used to show accurately what is happening in the full-scale unit. There are many examples where a pilot plant can be sensitized to behave like a full-scale plant by using non-geometric similarity techniques.

2.1. Exponential scale-up relationships

A generalized relationship can be constructed similar to that shown in Fig. 9. This shows a variety of published scale-up relationships in the area of liquid–solid suspension. There are scale-up proposals from Einenkel[8] which reference many other articles[9–17], particularly from the European literature. Another investigator, Herringe[20], found that scale-up relationships are a function of particle size. Data averaged for a variety of particle sizes will show scale-up parameters between the extremes found for small particles and large particles.

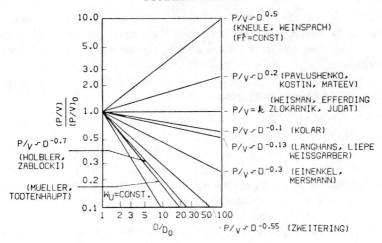

Figure 9   Illustration of the change of power per unit volume on scale-up for many different investigators in the area of solid suspension.

Articles by Rautzen et al.[19], and Connelly and Winter[25], suggest several exponential scale-up relationships for various types of processes. This chapter touches in more detail on solid–liquid mass transfer and solid–liquid suspension scale-up considerations.

All exponential relationships must be used with caution in practice for several reasons.

1. They are based on geometric similarity, which does not always keep all the relevant parameters increasing and decreasing as the need may be on scale-up.
2. They do not usually relate to important but somewhat complicated processes such as polymerization, crystallization and fermentation.
3. They often do not take into account scale-up effects such as differences in droplet or bubble sizes or linear velocities of immiscible fluids.
4. They do not apply to the macro scale, micro scale shear rate relationships which are important in many kinds of mixing processes.

In the remainder of this article, it will be attempted to give concepts and principles that can be used to apply qualitative tools to problems at hand. It usually turns out that estimates have to be made for several different important scale-up parameters, and a choice be made for scale-up for conservative yet practical combinations that will give full scale equipment with a realistic performance.

2.2. Fluid Motion on Scale-Up

Table 3 shows that impeller flow per unit volume normally decreases on scale-up. One investigator[21] used a radio transmitter in a small 8 mm diameter plastic pill and put a fixed ratio antenna around the impeller zone. The pill was circulated with the liquid in the tank and each time the pill came by the antenna a circulation was counted, in order to get an idea of circulation rates throughout the tank. As the tanks got bigger, particles tended to set up small, independent mixing zones along the way. The large tank has a longer blend time than that predicted by impeller pumping per unit volume. Table 3 shows at constant power per unit volume, where the scale ratio is 5:1, that the impeller pumping capacity per unit volume should be 1/3 of what it was in the small tank. The mean circulation time turns out to be more like 1/6, not 1/3, of the time in the small scale tank.

The 50-gallon tank had a single flow pattern. A large tank (1000 gals.) behaved as if it had two, five, or as many as 10 tanks in series, depending upon impeller speed. In a small tank, the ratio of impeller flow per unit volume approximately predicted the circulation time. The larger the tank size, the longer the circulation time becomes relative to tank volume/impeller flow rate.

An increase in the gas rate in a gas–liquid system also increases circulation time and the standard deviation of the blend time as well.

3. SCALE-UP CONCEPTS

Important for scale-up is a minimum size pilot plant for heterogeneous processes (see Fig. 10). Calculating the shear rate at the jet boundary gives a maximum shear rate of $10 \, sec^{-1}$. Across 1/8 of a centimeter, the shear rate is $9.5 \, sec^{-1}$; across a 1/4 centimeter, the shear rate is $7 \, sec^{-1}$; across half the impeller, there is a shear rate of $5 \, sec^{-1}$; and across the entire impeller, a shear rate of 0. A particle in the order of a centimeter or two in size is going to see a shear rate of 0 essentially, while a micron size particle will experience a shear rate of $10 \, sec^{-1}$.

The physical dimension of the impeller blade cannot get out of proportion to the particle size. In the plant, impellers are always

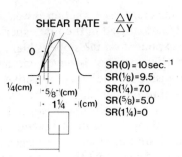

$$\text{SHEAR RATE} = \frac{\triangle V}{\triangle Y}$$

$SR(0) = 10\,\text{sec}^{-1}$
$SR(\tfrac{1}{8}) = 9.5$
$SR(\tfrac{1}{4}) = 7.0$
$SR(\tfrac{5}{8}) = 5.0$
$SR(1\tfrac{1}{4}) = 0$

Figure 10   Various shear rates around a radial flow impeller, blade width is 1.0 cm.

bigger than the particles dealt with. In the pilot plant, the general rule is the impeller blade width ought to be two or three times bigger than the particles dealt with in the process, or else shear rates on the particles will not be as expected.

There is no problem with scaling homogeneous chemical reactions. But if there are gas bubbles of 10 mm diameter, they cannot be dispersed the same way with a 5 mm blade height as with a 25 mm blade height. There is a different interaction of fluid shear rate with the bubbles. The results do not follow a logical scale-up relationship when the impeller is out of proportion to the bubbles. For example, in paper pulp, we find that we need a tank almost 60 cm in diameter to keep the ratio of fiber length and blade proportions within a reasonable range to make meaningful experiments for scale-up.

The major goal of a pilot plant or plant study is to find out the way the process responds to a change in mixing variables so the important parameters of the process can be indicated. This requires two things: (1) that there will be some data taken; (2) that they must be sufficient to describe important parameters to the mixing process result. Figure 11 shows the effect of a change in power on process result showing several different possibilities. Power is most conveniently changed by changing the speed of the impeller at constant diameter. When this is done the flow rate and shear rate are both changed, and if increases in both of these do not cause a change in process result there is little likelihood that other variables will. It is always possible, however, that an increase in flow rate might help, and an increase in shear rate might hurt the process, so they neutralize each other and there is little effect of power on the process. To be absolutely sure a study needs to be made with a different impeller diameter to see whether it was just a peculiarity of the particular combination used in the first experiment.

In Fig. 11, if the exponent on the slope is high, (A), that normally means a mass transfer process is involved, and quite typically gas–liquid mass transfer is the most sensitive to mixer power. Liquid–liquid mass transfer can also be involved, but liquid–solid mass transfer usually has a much lower slope. If the slope is zero, (E), it often is caused by a chemical reaction that is controlling which mixer variables are important. The jagged line shown on the left side of the curve indicates that the power levels are below those required to provide a satisfactory blend time in the tank and process results may be quite erratic. If the slopes are somewhere in the middle range, Curves (B) and (C), there is less of a clear-cut definition as to what the controlling variable may be, so either further data must be available or obtained or other things done in experimentation.

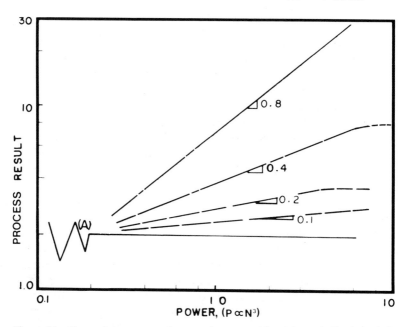

INTERPRETATION OF PILOT PLANT RESULTS
EFFECT OF POWER
ALL CONDITIONS CONSTANT EXCEPT POWER

Figure 11   Slope of process results vs. mixer power level is an indication of the controlling factor in the process.

Power per unit volume is a useful parameter, but it is a mistake to think that it is always going to be a constant (see Fig. 12). For some types of chemical reactions it is a constant because the micro scale mixing required is insensitive from whence the power came. Below the 500 micron level, it does not matter if the power came from a large or small impeller, so equal power per unit volume is a reasonable criterion for chemical reactions. Many times, in blending and slurry suspensions, the power per unit volume required drops on scale-up. However, to maintain equal blend time, power per unit volume increases with the square of the tank diameter.

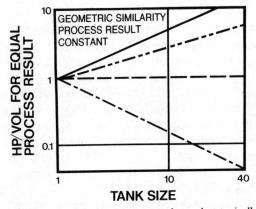

Figure 12 Power/volume can be a constant on scale-up, but typically it either goes up or it goes down. It is better used as a correlating parameter.

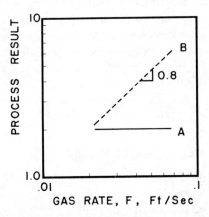

Figure 13 Two extremes of possible effects when gas rate is changed and a combination gas–liquid–solid slurry process curve "A" would be for liquid–solid controlled mass transfer and curve "B" would be for gas–liquid controlled mass transfer.

For a gas–liquid–solid process, then a simple measurement of the effect of gas rate can distinguish a liquid–solid controlled process from a gas–liquid controlled process. One does not change the mass transfer coefficient of a liquid–solid step by changing gas rates, so if there is a very low slope, liquid–solid mass transfer controls (see Fig. 13). On the other hand, if there is a big effect, then it is a gas–liquid mass transfer step that is involved.

4. SOME GENERAL PRINCIPLES OF PILOT PLANTING FOR SCALE-UP

As a general rule, vary the speed first. This raises the power level, which increases both flow and fluid shear. If that does not affect the process result, then there is little chance that any other mixing variable will. In order to find out whether flow or fluid shear is involved, then the D/T ratio should be examined. With various impellers in the tank one can see which D/T gives the best result.

To find out whether micro-scale mixing is more important than macro-scale, then systematic variation of the blade width in a series of experiments will vary the micro-scale mixing action separately from the macro-scale. This becomes more involved, but valuable results can be obtained.

Mixer manufacturers do not run a pilot plant on every application they analyze. Correlations and data are available to size about 95% of the applications encountered. Such processes as solids suspension and blending can be sized from existing correlations.

There are some processes where the scale-up effect is known, but quantitative values are needed for the small scale. If that is the case, all that is needed is to run one tank size. Examples of that would be gas–liquid mass transfer. If the process is so new that nothing is known about it, then normally two tank sizes should be used to estimate the scale-up relation empirically.

4.1. Gas–Liquid Mass Transfer Example

A typical gas–liquid mass transfer process usually involves something else besides gas–liquid dispersion. It may involve liquid and solids such as a biological step in a fermentation or there may be a chemical reaction occurring (Fig. 14). The procedure is to measure the mass transfer rate of the process and then estimate the gas–liquid concentration driving force and calculate a mass transfer

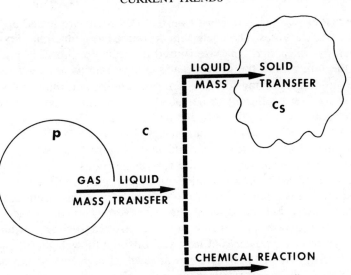

Figure 14   Gas–liquid mass transfer is usually in conjunction with liquid–solid mass transfer step or a chemical reaction.

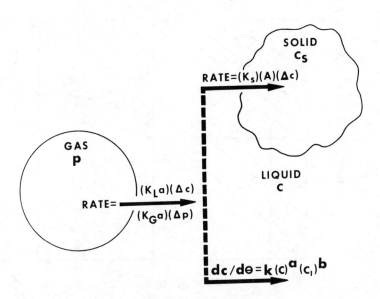

Figure 15   For scale-up purposes, the mass transfer coefficient is used rather than mass transfer rates.

coefficient (see Fig. 15). Mass transfer coefficients are scaled up, not mass transfer rates. The coefficient correlated as a function of power level, gas rates, and tank size is used for scale-up, shown in Fig. 16. In the larger tank the pressure and concentrations are used to get the gas–liquid driving force. The rate is known and that allows the mass transfer coefficient to be calculated. The mixer power level is determined as a function of gas rate. The key is that scale-up is not based on mass transfer rates but on mass transfer coefficients. A mixer can only affect mass transfer coefficients. Figure 17 shows a process curve for a specific tank size and various gas rates.

Figure 18 shows how sensitive gas–liquid mass transfer is to D/T ratio, which is related to the ratio of flow to fluid shear rate. At low mixer power (left-hand side) the gas overpowers the mixer and controls the flow pattern. A mixer controlled regime (center) is where the mixer power is at least two or three times higher than the gas energy. At the right is a power level so high that it makes no difference what kind of mixer we have.

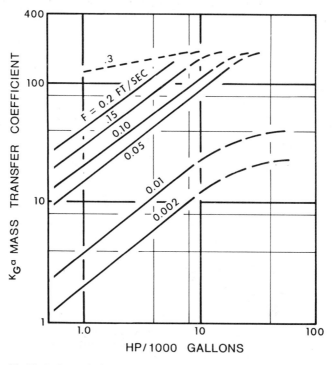

Figure 16   Typical correlation of gas–liquid mass transfer coefficient vs. power and superficial gas velocity.

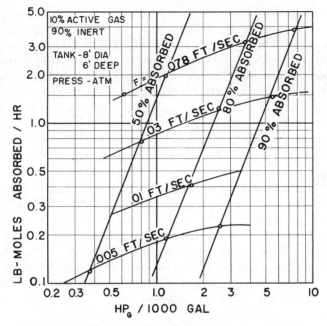

Figure 17 Typical correlation for an actual tank with a process fluid for various gas concentrations. This is derived from a mass transfer coefficient by putting in the actual driving forces for the process.

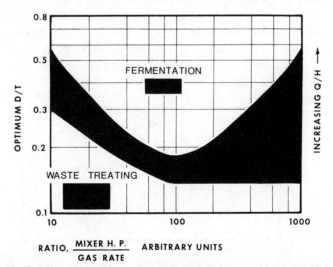

Figure 18 Relationship showing optimum D/T at various combinations of mixer HP to gas rate. On the left is gas controlled, in the center is mixer controlled and on the right mixer HP in large excess of gas HP.

In gas-controlled conditions a larger D/T is more effective. The optimum D/T is in the shaded band. In a mixer-controlled condition, a small D/T is best. Why is a D/T of 0.15 not used in a fermenter? D/T's of 0.33 to 0.44 are actually used. The reason is that those shear rates that are optimum for gas–liquid mass transfer would destroy the biological organisms, so there is no way to use small D/T's practically in a fermentation tank. Fermentation tanks, do not have the optimum liquid–gas mass transfer because to do that would require geometry which would destroy the biological process. One cannot, as a rule, optimize every individual mixing step in the process.

Why are not large D/T's used in waste treating tanks? The tanks are so large, one cannot practically put in a 0.33 D/T, so they are not designed for the optimum mass transfer, but are designed for practical mass transfer and acceptable economics.

Chapter 5 by Smith discusses the mechanics of bubble dispersion and its effect on power draw and mass transfer. In general, power per unit volume approximately correlates the mass transfer coefficient of widely different impeller types.

4.2. Fermentation Pilot Plant Example

Consider first a batch fermentation with a standard D/T ratio of 1/4 and a blade width to impeller diameter ratio that is normal, typically 1/5. That impeller gives a shear rate of 1 relative to other designs. The yield at the end of five days increases with increased power to a certain level and then falls off with further increase (see Fig. 19). The suspicion is that too high a shear rate is affecting the organisms. To be on the safe side, a reasonable yield is chosen.

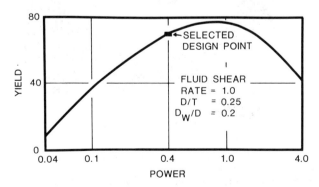

Figure 19   Typical data from fermentation process showing that as power increases, yield after a 5-day batch fermentation reaches a maximum and then will decrease.

For the pilot plant the variable levels chosen must cause fermentation with the gas–liquid mass transfer rate which delivers oxygen to the liquid. The liquid–solid mass transfer rate must be considered, because it supplies oxygen to the organism, and the shear rate around the impeller is important.

The first scale-up calculation, Table 6, uses geometric similarity with the same D/T ratio, although the tank is going to be taller in proportion, since it has a Z/T ratio now of about 2:1 rather than 1:1. For equal gas–liquid mass transfer, and geometric similarity, the tip speed which relates to the maximum shear rate, has gone up 80%. Another mixer with a larger D/T ratio of 0.38 is then considered. To get the same mass transfer rate, a slower speed with the bigger impeller is necessary. That reduces the maximum impeller zone shear rate to 40% higher than it was in the pilot plant.

The biggest item in the cost of a mixer is the torque, so by cutting the speed in half, the torque has been doubled. That means the mixer is going to cost more, almost double the cost of the mixer in Column B. Using a 0.6 D/T ratio the shear rate can be reduced to what it was in the pilot plant. By doing so, four times more torque is needed and the mixer will cost four times more than that in Column B.

More pilot plant experiments may be necessary to see if mixer "B" or "C" will be satisfactory. An impeller 1/3 the blade width at the same D/T, will run faster and will develop 40% higher shear than the original impeller (see Fig. 20). We get similar data at low power levels, but now at the design point the yield curve starts

Table 6   Relationship between Process Performance and Cost for Three Different D/T Ratios Selected for Full Scale Mixers.

	Pilot A	Plant B	Plant C	Plant D
T	1.2	7.0	7.0	7.0
P	1.0	340	340	340
N	1.0	0.31	0.16	0.075
D/T	0.25	0.25	0.38	0.6
No. of Imp.	1	2	2	2
Z	1.2	12.0	12.0	12.0
F	F	10 F	10 F	10 F
Max. I.Z.	1.0	1.8	1.4	1.0
Torque	—	1.0	2.0	4.1
cost ($)		10,000	18,000	35,000

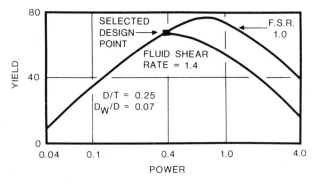

Figure 20   A second impeller type, having a shear rate 1.4 × original impeller, shows a decrease in yield at a lower power level.

to flatten out. This is a good indication the shear rate is what's causing the yield increase.

Mixer "C" is a good choice. In order to see if mixer "B" is satisfactory, the blade width could be reduced further by another factor of 1/3. However, then the blade is smaller than the gas bubbles in the system. The blade width should be bigger than the particles to be dispersed, so there is no way to find out for sure if mixer "B" will work in this size of pilot tank. A larger pilot plant must be built where blade proportion would be suitable.

4.3. Typical Fermentation Specifications

There are four ways in which mixers are often specified when considering putting larger units in a fermentation plant. In this context either a larger tank with a suitable mixer or improvement of the productivity of a given tank by installation of a different combination of mixer horsepower and gas rate are possibilities. The four bases for specification are listed below:

1. Change in productivity requirements based on production data with a particular size fermenter in the plant.
2. New production capacity based on pilot plant studies.
3. Specification of agitator based on the sulfite absorption rate in aqueous sodium sulfite solution.
4. Specification of the oxidation uptake rate in the actual broth for the new system.

The power input from the gas will increase on scale-up. This is because there is a greater head pressure on the system, and there is also an increasing gas velocity. It may be that the power level for the mixer may be reduced in order to maintain a particular mass transfer coefficient, K_Ga, since the energy from the gas going through the tank is higher.

However, this lower power level may reduce the liquid–solid mass transfer rate. The blend time will also be increased, so that this reduced power level is not usually practical.

4.4. Scale-Up Based on Data from an Existing Production Plant

If data are available on a fermentation in a production-size tank, scale-up may be made by increasing, in a relative proportion, the various mass transfer, blending and shear rate requirements for the full-scale system. For example, it may be determined that the new production system is to have a new mass transfer which is some fraction of the existing mass transfer rate. There may be specifications put on maximum or average shear rates, and there may be a desire to look at changes in blend time and circulation time. In addition, there may be a desire to look at the relative change in the CO_2 stripping efficiency in the revised system.

At this point, there is no reason not to consider any size or shape of tank. Past tradition for tall, thin tanks, or short, squat tanks, or elongated horizontal, cylindrical tanks does not mean that those traditions must be followed in the future. Figures 14 and 15 illustrate the principle involved in the gas–liquid mass transfer. There are three different mass transfer and reaction steps commonly present in fermentation. The mass transfer rate must be divided by a suitable driving force, which yields the mass transfer coefficient required. The mass transfer coefficient is then scaled to the larger tank size and is normally related to superficial gas velocity to an exponent, power per unit volume to an exponent, and to other geometric variables such as the D/T ratio of the impeller.

A thorough analysis involves every proposed tank shape, the gas rate range required, the gas phase mass transfer driving force, and then the required K_Ga to meet that. Reference is made to mass transfer rate for the mixer under the condition specified, to obtain the right mixer power level for each gas rate and at each D/T ratio.

At this point, the role of viscosity must be considered. Figure 21 gives typical data on the effect of viscosity on mass transfer coefficient[4]. It is necessary to measure viscosity with a viscometer which mixes while it measures viscosity. Figure 22 illustrates the

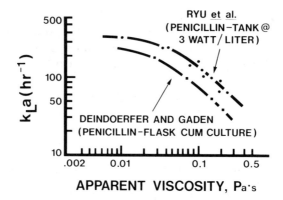

Figure 21   Illustration that mass transfer coefficient decreases with the viscosity of medium in fermentation.

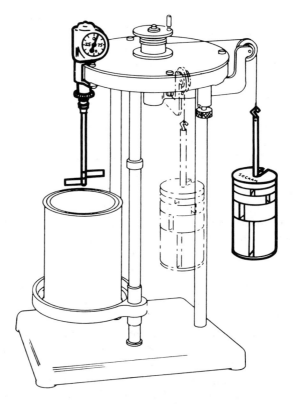

Figure 22   Stormer viscometer used for measuring the viscosity of 2-phase systems.

Stormer viscometer which is one device that can be used to establish viscosity under mixing conditions with known shear rates.

For the new tank size, estimates should be made of the shear rate profile around the system. Then using the relationship that viscosity is a function of shear rate, and the fact that it is shear stress [Shear Stress = μ (Shear Rate)], that actually carries out the process, the viscosity throughout the tank as well as the product of viscosity and shear rate can be estimated. Estimates can then be made of how different the proposed new tank may be compared to the existing known performance of the production tank.

4.5. Data Based on Pilot Plant Work

To keep ratios of impellers, gas bubbles and solid clumps in the fermentation related to full scale, the impeller size and blade width in the small scale must always have a physical dimension two or three times bigger than the particle size under concern. It is possible to model the fermentation biological process from a fluid mechanics standpoint, even though the impeller is not related properly geometrically to the gas–liquid mass transfer step. Thus, one size of pilot plant might be usable for evaluating one or two of the fermentation mass transfer steps, and/or chemical reaction steps, but might not be suitable for analysis of other mass transfer steps. The decision then is based on how suitable existing data are for any steps which are not modeled properly in the pilot plant.

Ideally, data should be taken during the course of the fermentation of gas rate, gas absorption, dissolved oxygen level, dissolved carbon dioxide level, yield of desired product and other parameters which might influence the decision on the overall process. Figure 23 shows a typical set of data for this situation.

If the pilot plant is to duplicate certain properties of fluid mixing, then it may be necessary to use non-geometric impeller and tank geometry scale-up to duplicate mixing performance and not geometric similarity. As a general rule, geometric similarity does not control any mixing scale-up property whatsoever.

It may also not be possible to duplicate all of the desired variables in each run, so a series of runs may be required changing various relationships systematically and then a synthesis made of the overall results.

One variable in particular is important. The linear superficial gas velocity should be run in a few cases at the levels expected in the full scale plant. This means that foaming conditions will be more

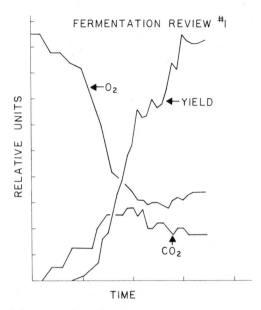

Figure 23   Typical relationship of dissolved oxygen, dissolved CO_2, and fermentation yield for batch fermentation.

typical of what is going to happen in the plant and the pilot fermenter should always be provided with enough head space to make sure the foam levels can be adequately controlled in the pilot plant. As a general rule, foam level is related to the square root of the tank diameter on scale-up or scale-down.

If it is desired to duplicate maximum impeller zone shear rates on a small scale, there may be a very severe design problem in the mechanics of the pilot plant mixer, such as the shaft speed, mechanical seals, and other items. This means that careful consideration must be given to the type of runs to be made and whether the design of the pilot plant equipment must duplicate the maximum impeller zone shear rates in the plant, and whether this type of data will be obtained on a separate pilot unit dedicated to study that particular variable.

Figure 24 shows what often happens in terms of correlating mass transfer coefficient $K_G a$, with power and gas rate in the pilot plant. This curve is then translated to a suitable relationship for full scale. It is possible to consider that with the higher superficial gas velocity, the power level may be reduced in the full scale to keep the same mass transfer coefficient. The box on the right in Fig. 24 (Pilot Scale)

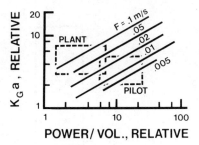

Figure 24   Correlation of K_Ga vs. power per unit volume and gas flow in pilot plant and plant size units.

shifts to the box on the left (Plant Scale). This should be considered, but it should be borne in mind that this changes the ratio of the mixer power to gas power level in the system; changes the blend time; changes the flow pattern in the system; changes the foaming characteristics and also can markedly affect the liquid–solid mass transfer rate, Fig. 25, if that is important in the process.

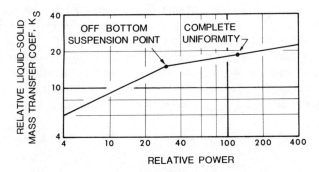

Figure 25   Illustration that the effect of power on the liquid–solid mass transfer coefficient in liquid–solid systems has more of an effect up to the off-bottom suspension point than it does up to the uniformity point.

In all cases, a suitable mass transfer driving force must be used. Figure 26 illustrates a typical case for fermentation processes, and illustrates that there is a marked difference between the arithmetic average driving force, the log-mean driving force, and the exit gas driving force. In a larger fermenter, gas concentrations are essentially step-wise staged function, and a log-mean average driving force has been the most fruitful. Figure 27 illustrates a small laboratory fermenter with a Z/T ratio of 1, and in this case depend-

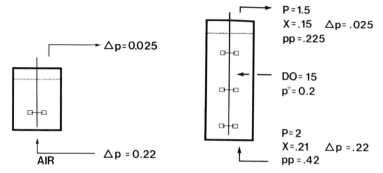

Figure 26 Gas concentration driving force for well mixed batch fermenter, indicating that a choice must be made between an average driving force and an exit driving force.

Figure 27 On large scale fermenter, the height and mixing intensity is such that log-mean driving force is more suitable than exit driving force.

ing on the power level, an estimate must be made of the gas mixing characteristics and an evaluation made of the suitability of the exit gas concentration for the driving force compared to the log-mean driving force. This is one area which needs to be explored in the pilot program and the calculation procedures.

Just to indicate another peculiarity, in the waste treating industry, it is quite common to run an unsteady state re-aeration test. The tank is stripped of oxygen; air is introduced with the mixer operating and the dissolved oxygen level increase is monitored up to where

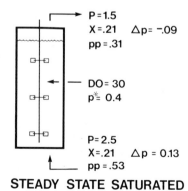

STEADY STATE SATURATED

Figure 28 In an unsteady state re-aeration test, process equilibrium means that the top of the tank has a negative driving force while the bottom of the tank has a positive driving force.

the tank is saturated and no further mass transfer occurs. At that point, the dissolved oxygen level is usually between the saturation value at the top and the saturation value at the tank bottom pressure (Fig. 28). This means that for steady state there must be absorption in the bottom and stripping at the top, and a very untypical mass transfer situation results compared to what happens on a waste treatment full scale aerator. Running experimental tests and basing a lot of calculations on that particular driving force gives a marked difference in the estimate of the way the mixer will operate in practice.

4.6. Sulfite Oxidation Data

Excess sodium sulfite with suitable catalysts which can be used will keep the dissolved oxygen level at zero throughout the batch run, and data have been obtained on small and large size fermentation tanks with this system. One caution is that the data should have been taken when the tank is completely clean of anti-foam chemical which may be residual from the fermentation process. These anti-foam chemicals can cause marked variations in the mass transfer coefficient.

A relationship between the sulfite oxidation number and the performance required in the fermenter gives a perfectly valid way to specify equipment, and tests can be run to give an indication of the overall mass transfer rate ensuing.

4.7. Oxygen Uptake Rate in the Broth

If it is desired to relate fermenter performance to oxygen uptake rate in the broth, this number can be specified along with suitable desired gas rates and the mixer estimated based on this performance. Again, the link between this particular mass transfer specification and the actual performance of the fermentation must be known[5,6]. When based on pilot plant data, then the effect of the different shear rates, different blend times on both the mass transfer relationship, viscosity and the resulting fermentation must be considered.

4.8. Zinc Purification Process

In a process study made in conjunction with American Zinc[22], an illustration of interaction of many variables in the process can be shown. In this particular process, it was desired to retain the cadmium at a high level in the purification process and to remove cobalt and arsenic. The process involved adding zinc dust to a zinc solution from a leach system. There was usually a sudden drop in the cadmium as well as the cobalt and arsenic, and then if sufficient time was possible, the cadmium was liberated in the process and came back to acceptable recovery level.

It was desired to see whether mechanical mixing could affect the time of this reaction and also affect the recovery in the cadmium and the rejection of the arsenic and cobalt.

Figure 29 shows the results in a 1016 mm diameter pilot plant study. In this study for any given run there were typical curves shown in Fig. 29, and at a particular optimum horsepower level there was a good recovery of the desired cobalt in the process. On that same scale, Fig. 30 shows that the 355 mm diameter impeller was much more effective than the 230 mm impeller indicating that flow-to-shear ratios are important.

Data were also obtained on a 560 mm diameter scale and while similar process profiles were obtained with power and time, it turned out that the two impeller diameters tested, 122 mm and 203 mm, gave the same result (Fig. 30).

The interpretation here is that on the small scale, 560 mm, there was so much extra pumping capacity and a short blend time compared to 1016 mm diameter tank, and of the full-scale tank in the plant, that any reduction in the pumping capacity due to the 122 mm diameter impeller was not sufficient to affect the process result.

On the other hand, on the 1016 mm scale any reduction in the

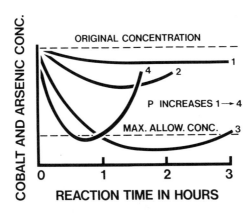

Figure 29   Illustration of four different power levels in zinc purification process described in text.

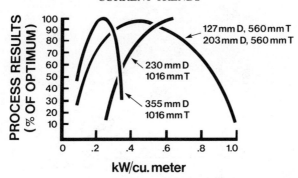

Figure 30   Illustration that impeller diameter makes a difference on small pilot tank while impeller diameter does not make a difference on small diameter tanks.

pumping capacity and increase in shear did affect the process and it would be expected that D/T would be an important ratio in going up to a full size system.

If it were desired to obtain this kind of information on only the 560 mm diameter scale, it would have been necessary to drastically change the impeller blade width ratio to markedly cut the pumping capacity down into the range where it would be in the full-scale plant to show whether this pumping capacity in full scale will be a serious detriment to the process performance.

Plant scale data reported indicate that by using suitable D/T ratios in the plantt, plant scale performance was predicted by looking at the overall scale-up effect between the 1016 mm diameter tanks, and taking into account the optimum D/T ratio needed to carry out this process.

4.9.   Flocculation

Flocculation is a process which is extremely sensitive to mixing variables. In a series of publications, Oldshue and Mady[23, 24] showed that for two different impellers the G factor for the minimum turbidity, which is a measure of optimum flocculation performance decreased with tank diameter as shown in Fig. 30. The G factor is the square root of the power per unit volume divided by viscosity. The study found that this minimum value of turbidity, which corresponds to optimum process performance, occurred at a particular impeller speed; speeds above and below this gave poor performance. Flocculation particles are increased in tank size by fluid shear rate which gets particles from various parts of the tank together, but the particle

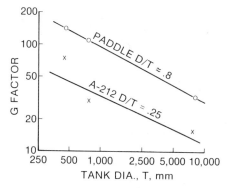

Figure 31   Results of flocculation experiments on small and large tanks as described in text.

size is reduced by the shear stress which results so there is an optimum speed for every particular geometry.

Another observation about Fig. 31 that there is a decreasing power per unit volume required as tank size increases to obtain this optimum flocculation performance.

5. SUMMARY

The pilot plant has a much higher pumping capacity, a lower average impeller macro scale shear rate and a higher maximum macro scale impeller shear rate. To change those relationships to be more representative of the full scale tank requires a change in geometry away from geometric similarity.

If one desires to optimize the pilot plant results using a standard geometry there, the consequences of changes in fluid mixing parameters must be carefully considered on scale-up. On the other hand, if one wishes to duplicate plant scale performance as much as possible in the pilot plant, then the geometry of the pilot plant must be made different from the plant to approach the required fluid mixing parameters.

6. NOTATION

P power
N impeller speed
D impeller diameter

Z liquid level
T tank diameter
m mass of fluid
a acceleration of fluid

Greek Symbols

μ viscosity
ρ surface tension
g gravitational constant
N_{Re} Reynolds number
N_{Fr} Froude number
N_{We} Weber number
$K_G a$ gas–liquid mass transfer coefficient
$K_L a$ liquid–liquid mass transfer coefficient
K_S liquid–solid coefficient
D.O.   dissolved oxygen
P^* equilibrium partial pressure in gas phase corresponding to dissolved solute concentration

REFERENCES

1. Oldshue, J. Y. (1970). *Spectrum of Fluid Shear Rates in a Mixing Vessel*, CHEMECA '70, Australia (Butterworth).
2. Oldshue, J. Y. (1966). Fermentation mixing scale-up techniques, *Biotechnol. Bioeng.*, **VIII**, 3–24.
3. Oldshue, J. Y., and Gretton, A. T. (1954). *Chem. Eng. Prog.*, **50**, 615.
4. Deindoerfer, F. H., and Gaden, E. L. (1955). *Appl. Microbiol.*, **3**, 253.
5. Oldshue, J. Y., Coyle, C. K., et al. (1977). Fluid mixing variables in the optimization of fermentation production, *Process Biochem.*, **13**(11), (Published Wheatland Journals Ltd., Watford, Herts, England).
6. Ryu, D. Y., and Oldshue, J. Y. (1977). A reassessment of mixing cost in fermentation processes, *Biotechnol. Bioeng.*, **XIX**, 621–629.
7. Levenspiel, Octave, and Soon J. Kang, (1976). *Chem. Eng.*, **83**, 141.
8. Einenkel, W. D. (1980). German *Chem. Eng.*, **3**, 118–124.
9. Kneule, F., and Weinspach, P. M. (1967). *Verfahrenstechnik (Mainz)* **1**(12), 531–540.
10. Zweitering, Th. N. (1958). *Chem. Eng. Sci.*, **8**, 244–253.
11. Pavlushenko, J. S., and Kostin, N. M. (1957). *J. Appl. Chem.*, *USSR*, **30**, 1235–1243.
12. Kolar, V. (1961). *Coll. Czech. Chem. Commun.*, **26**, 613–627.
13. Weisman, J., and Efferding, L. E. (1960). *AIChE J.*, **6**(3), 419–426.
14. Einenkel, W. D., and Mersmann, A. (1977). *Verfahrenstechnik (Mainz)*, **11**(2), 90–94.
15. Kotzek, R., Langhans, G., Liepe, F., and Weissgarber, H. (1969). *Mitt. Inst. Chemieanl.*, **9**(2), 53–58.
16. Zlokarnik, M., and Judat, H. (1969). *Chem. Ing. Tech.*, **41**(23), 1270–1273.
17. Holbler, T., and Zablocki, J. (1966). *Chem. Tech. (Leipzig)*, **18**(11), 650–652.

18. Mueller, W., Todtenhaupt, E. K. (1972). *Aufbereit. Tech.* **13**(1), 38–42.
19. Rautzen, R. R., Corpstein, R. R., and Dickey, D. S. (1976). *Chem. Eng.*, 119–126.
20. Herringe, R. A. (1979). *Proc. 3rd European Conf. on Mixing*, York, England, April, Vol. 1, pp. 199–216 Paper D1.
21. Middleton, J. A. (1979). *Proc. 3rd European Conf. on Mixing*, York, England, Vol. 1, p. 15, Paper A2.
22. Carpenter, R. K., and Painter, L. A. (1955). (Presented 1955 Annual Meeting AIMME.)
23. Oldshue, J. Y., and Mady, O. B. (1978). *Chem. Eng. Prog.*, **74**, 103.
24. Oldshue, J. Y., and Mady, O. B. (1979). *Chem. Eng. Prog.*, **75**, 72.
25. Connelly, J. R., and Winter, R. L. (1969). *Chem. Eng. Prog.* **65**(8), 70.

Subject Index